Nuclear Power Plants - Processes in the Nuclear Fuel Cycle

Edited by Nasser Awwad

Published in London, United Kingdom

IntechOpen

Supporting open minds since 2005

Nuclear Power Plants - Processes in the Nuclear Fuel Cycle
http://dx.doi.org/10.5772/intechopen.87697
Edited by Nasser Awwad

Contributors
Madalin Ion Rusu, Valeriu Savu, Dan Savastru, Victor Kozlov, Yasuo Hirose, Mostafa Esmaeili Shayan, Farzaneh Ghasemzadeh, Akbar Abbasi, Andre S. De Aguiar, S.M Lee, Gaianê Sabundjian, Mustapha Makhloul, Hamid Boukhal, Tarek El Bardouni, El Mahjoub Chakir, Mohamed Kaddour, Sanae Elouahdani, Nasser S Awwad

First published in London, United Kingdom, 2021 by IntechOpen
IntechOpen is the global imprint of INTECHOPEN LIMITED, registered in England and Wales, registration number: 11086078, 5 Princes Gate Court, London, SW7 2QJ, United Kingdom
Printed in Croatia

British Library Cataloguing-in-Publication Data
A catalogue record for this book is available from the British Library

Additional hard and PDF copies can be obtained from orders@intechopen.com

Nuclear Power Plants - Processes in the Nuclear Fuel Cycle
Edited by Nasser Awwad
p. cm.
Print ISBN 978-1-83968-330-5
Online ISBN 978-1-83968-331-2
eBook (PDF) ISBN 978-1-83968-335-0

For EU product safety concerns:
IN TECH d.o.o., Prolaz Marije Krucifikse Kozulić 3, 51000 Rijeka, Croatia, info@intechopen.com or visit our website at intechopen.com.

Meet the editor

Dr. Nasser Awwad received his Ph.D. in inorganic and radiochemistry in 2000 from Ain Shams University and his Ph.D. at Sandia National Labs, New Mexico, USA, 2004. Nasser Awwad was an Associate Professor of Radiochemistry in 2006 and Professor of Inorganic and Radiochemistry in 2011. He has been a Professor at King Khalid University, Abha, KSA, from 2011 until now. He has published two chapters in the following books "Natural Gas - Extraction to End Use" and "Advances in Petrochemicals". Pro Awwad has edited four books (Uranium, New trends in Nuclear Sciences, Lanthanides, and Nuclear Power Plants) and he has co-edited two books ("Chemistry and Technology of Natural and Synthetic Dyes and Pigments" and "Chromatography - Separation, Identification, and Purification Analysis"). He has also published 95 papers in ISI journals. He has supervised 4 Ph.D. and 18 MSc students in the field of radioactive and wastewater treatment. He has participated in 26 international conferences in South Korea, the USA, Lebanon, KSA, and Egypt. He has reviewed 2 Ph.D. and 13 MSc theses. He participated in 6 big projects with KACST at KSA and Sandia National Labs in the USA. He is a member of the Arab Society of Forensic Sciences and Forensic Medicine. He is a permanent member of the American Chemical Society, and a rapporteur of the Permanent Committee for Nuclear and Radiological Protection at KKU. He is Head of the Scientific Research and International Cooperation Unit, Faculty of Science, King Khalid University.

Contents

Preface

This book focuses on the hot topics related to nuclear power plants. The first of these hot topics is how to use nuclear power plants to generate electric energy using nuclear fuel such as Uranium Oxide (UOX). The second hot topic provides an advanced level of understanding of the complete fuel cycle by following nuclear fuel from its origin and fabrication, through its stay in the reactor with all alterations induced there, and ending with reprocessing options and waste management issues for the spent nuclear fuel. In other words; "Nuclear Power Plants - Processes in the Nuclear Fuel Cycle". Moreover, this book covers radiation protection issues throughout the nuclear fuel cycle. There are two types of radiation sources: naturally occurring radioactive materials (NORM) and technologically enhanced naturally occurring radioactive materials (TENORM), which are considered a raw material in the nuclear industry. Safety issues in nuclear power plants and radioactive wastes will be discussed. In addition, the future of nuclear power, nuclear fuel, uranium, and the advantages of nuclear energy will be covered. In addition, the reader of the book will find the answer to the important question, is nuclear energy a renewable or nonrenewable source?

Nasser S. Awwad, Ph.D.
Professor of Inorganic and Radiochemistry,
Chemistry Department,
Faculty of Science,
King Khalid University,
Saudi Arabia

Section 1

Nuclear and Solar Energy

Chapter 1

Introductory Chapter: From the Cradle to the Grave for the Nuclear Fuel Cycle

Nasser S. Awwad

1. Introduction

This chapter will focuses on the brief topics of Nuclear Fuel Cycle. It will provide advanced level for understanding of the complete fuel cycle by following nuclear fuel from its origin and fabrication, through its stay in the reactor with all alterations induced there, and ending with reprocessing options and waste management issues for the spent nuclear fuel. In other words: "Nuclear fuel – from cradle to grave". Moreover, it covers radiation protection issues throughout the nuclear fuel cycle.

2. Front end of the fuel cycle which includes the following items

2.1 Uranium exploration and mining

Uranium mining is the process by which uranium metal is extracted from the earth. In 2019, the worldwide generation of uranium produced 53,656 tons. The top three producers were Kazakhstan, Canada and Australia, which together account for 68 percent of the world's uranium generation. Namibia, Niger, Russia, Uzbekistan and China were other vital uranium-producing nations in excess of 1,000 tons per year [1]. Mining uranium is almost exclusively used as fuel for nuclear power plants. Uranium is extracted by in-situ filtration (57% of the world's generation) or by ordinary underground or open-pit metal mining (43% of the generation). A filtering arrangement is pumped down in the in-situ mining, penetrating holes into the uranium metal store where the mineral minerals are broken up.

At that point, the uranium rich liquid is pumped back to the surface and prepared to extract the uranium compounds from the structure in regular mining, metals are treated by pounding the metal materials to a uniform molecule measure and after that treating the mineral to extricate the uranium by chemical leaching [2]. The processing handle usually produces dry powder-form fabric composed of common uranium, "yellowcake," which is marketed as U3O8 on the uranium display.

2.2 Uranium enrichment

The nuclear fuel used in a nuclear reactor needs to have a higher U235 isotope concentration than that found in natural uranium ore. In light water reactors (the most common reactor design in the USA), U235 is fissionable when concentrated (or 'enriched'). The nucleus of the atom breaks apart during fission, creating both heat and extra neutrons.

These extra neutrons can cause additional, nearby atoms to fission under controlled conditions and a nuclear reaction can be maintained. Via a controlled nuclear reaction inside the nuclear reactor, the heat energy released can be harnessed to generate electricity. The U235 isotope is commercially enriched to 3 to 5 percent (from the natural state of 0.7 percent) and then further processed for nuclear fuel output.

Uranium oxide is converted to the chemical form of uranium hexafluoride (UF6) at the conversion plant to be used in an enrichment facility. UF6 is used for a few reasons; 1) the fluorine portion has only one isotope that occurs naturally, which is an advantage during the enrichment processing the fluorine does not contribute to the weight difference when separating U235 from U238), and 2) UF6 exists as a gas at an optimal operating temperature. There are several enrichment processes utilized worldwide. They are:

- Gaseous Diffusion
- Gas Centrifuge
- Laser Separation

The first industrial method used to enrich uranium in the United States was gaseous diffusion. These facilities used large quantities of energy and the existing gaseous diffusion plants became outdated as the centrifuge technology matured. All of them have been replaced worldwide by second-generation technology, which needs much less energy to generate comparable quantities of separated uranium. Uranium hexafluoride (UF6) gas was injected into the pipes of the plant at a gaseous diffusion enrichment plant, where it was pumped through special filters called barriers or porous membranes.

The holes in the barriers were so small that the UF6 gas molecules barely had enough space to move through. The isotope enrichment occurred because the lighter molecules of UF6 gas (with atoms U234 and U235) spread faster through the barriers than the heavier molecules of UF6 gas containing U238.

However one barrier wasn't enough. Until UF6 gas contained enough U235 to be used in nuclear fuel, several hundreds of barriers were required, one after the other. The enriched UF6 gas was separated from the pipes at the end of the operation and condensed back into a liquid that was then poured into containers. Before it was transported to fuel fabrication facilities, the UF6 was allowed to cool and solidify. The current method by which commercial enrichment is conducted in the United States is gas centrifuge enrichment. UF6 gas is positioned and rotated at a high velocity in a gas centrifuge cylinder. A strong centrifugal force is generated by this rotation so that the heavier gas molecules (UF6 containing U238 atoms) travel towards the outside of the cylinder. Closer to the middle, the lighter gas molecules (containing U235) gather. The slightly enriched stream in U235 is extracted and fed into the next centrifuge; the next higher level. The slightly depleted stream is recycled back into the next lower stage (with a lower concentration of U235).

Long lines of several revolving cylinders are contained in a gas centrifuge factory. In both series and parallel formations, those cylinders are related. To shape trains and cascades, centrifuge machines are interconnected. The UF6 is enriched to the required amount at the final withdrawal stage.

For potential use to enrich uranium, the laser separation technology is under development. By extracting isotopes of uranium with lasers, uranium can be enriched. Laser light can excite molecules; this is called photoexcitation. Lasers will raise the energy in the electrons of a particular isotope, alter its properties and allow it to be separated. Three main systems, which are laser systems, optical systems, and the separation module system, are used in the enrichment process. In order to deliver highly monochromatic light, tunable lasers can be produced. A particular isotopic species may be photo-ionized by the light from these lasers while not affecting other isotopic species. Chemically, the infected species is then modified, which allows the substance to be isolated. The laser separation technology developed by DOE uses as its feed material a uranium metal alloy, while UF6 is used as the feed material in the Separation of Isotopes by Laser Excitation process.

2.3 Fuel fabrication

Fuel production installations turn enriched uranium into nuclear reactor fuel. Mixed oxide (MOX) fuel, which is a mixture of uranium and plutonium, may also be used in fabrication. Usually, the manufacture of fuel for light water reactors (LWR) (regular commercial power reactors) begins with the receipt of low-enriched uranium from an enrichment facility, in the chemical form of uranium hexafluoride (UF6).UF6 is heated to a gaseous state in solid form in tubes, and then the UF6 gas is chemically treated to form a powder of uranium dioxide (UO2). This powder is then pressed into pellets, loaded into Zircaloy tubes, sintered into ceramic form, and constructed into fuel assemblies. If it is a boiling-water reactor or a pressurized-water reactor depends on the type of light water reactor.

MOX fuel differs from low-enriched uranium fuel in that the powder used to form fuel pellets is comprised of both uranium dioxide (UO2) and plutonium dioxide (PuO2). The MOX fuel will be used in light-water reactors.

Small reactors that do not produce electrical power but are used for research, testing and training are non-power reactors. Analysis reactors and reactors used to manufacture irradiated target materials may be included in non-power reactors. The configuration of the fuel varies with the kind and manufacturer of the reactor. The plate-type fuel consists of several thin sheets containing an aluminum-clad uranium mixture. Another fuel is in the form of rods and is made up of a combination of uranium and zirconium/hydride.

3. Nuclear fuel in reactor

Nuclear fuel is the fuel which is used to support a nuclear chain reaction in a nuclear reactor. These fuels are fissile, and the radioactive metals uranium-235 and plutonium-239 are the most common nuclear fuels. A cycle known as the nuclear fuel cycle is made up of all the steps involved in collecting, refining, and using this fuel. Many nuclear fuels produce heavy elements of fissile actinide that are able to undergo and sustain nuclear fission. Uranium-233, uranium-235 and plutonium-239 are the three most applicable fissile isotopes. As a slow-moving neutron strikes the unstable nuclei of these atoms, they split, forming two daughter nuclei and two or three more neutrons. Then these neutrons go on to split more nuclei. This produces a self-sustaining chain reaction that is regulated or uncontrolled by a nuclear bomb in a nuclear reactor.

4. Back end of the fuel cycle

For a period of five years or so, the fuel is first placed in a storage pool, the time to let the most active fission products decrease or vanish. After those five years, a decision as to whether or not to reprocess is made. If not, so as it is the fuel must be stockpiled.

Research is currently underway on the feasibility of the final disposal of spent fuel deep underground; decisions on such disposals are yet to be made. Meanwhile the waste accumulated above ground is accumulating around the power plants. If a reprocessing decision is taken, the spent fuel is transported to a reprocessing plant where it is deposited in a nearby pool for a few more years. Reprocessing requires separating what can be recycled from what can be labeled as actual waste - uranium and plutonium. Usually, 95% of the fuel consists of plutonium, which also contains about 1% of the fissile isotope 235, more than natural uranium. The spent fuel also contains an additional 1% of plutonium, of which 70% of the isotopes are fissile and can generate electricity. It is possible to re-enrich this uranium and recycle the plutonium to join the fresh fuel composition to power other reactors.

Fission products and small actinides make up the remaining 4 percent of the spent fuel. They account for about 98% of their gamma and beta radioactivity. These are the real products of waste. This waste is highly radioactive, but it is conditioned by embedding it in glasses or ceramics that provide fewer long- term environmental risks than the disposal without reprocessing of spent fuel 'in-state.' The final disposition of these vitrified waste is yet to be determined, but their stockpiling in interim storage facilities is less of a concern as their mass is much smaller than the spent fuel.

Over a number of years, the IAEA has developed a comprehensive set of safety series documentation, which addresses, in a structured manner, many of the various nuclear fuel cycle safety needs identified by Member States. Since 1996 the IAEA Safety Standards series of documents has been subject to a process of planned change from its original structure of Safety Fundamentals, Safety Standards, Safety Guides and Safety Practices, to a new structure with a single Safety Fundamentals document supported by Safety Requirements and Safety Guides. The existing IAEA documents cover the safety of nuclear installations (predominantly, but not exclusively, nuclear power plants), radioactive waste management, radiation protection and transport safety [3].

5. Nuclear Power Station at production of energy

In the 1950s, the first commercial nuclear power plants began operation. Out of about 440 power plants, nuclear energy now generates about 10 percent of the world's electricity. Nuclear power is the world's second largest low-carbon power source (29 percent of the total in 2017). One of condition to be as source of renewable energy. Over 50 countries utilize nuclear energy in about 220 research reactors. In addition to research, these reactors are used for the production of medical and industrial isotopes, as well as for training. In 2018, 12 countries generated at least one quarter of their electricity from nuclear power. About three-quarters of France's energy comes from nuclear power, more than half from Hungary, Slovakia and Ukraine, and one-third or more from Belgium, Sweden, Slovenia, Bulgaria, Switzerland, Finland and the Czech Republic.

Normally, South Korea gets more than 30 percent of its electricity from nuclear power, while about one-fifth of its electricity comes from nuclear power in the USA, UK, Spain, Romania and Russia. For more than one-quarter of its energy, Japan is used to rely on nuclear power and is expected to return to somewhere near that amount.

With a total net capacity of 1.6 GWe, Mexico has two operable nuclear reactors. In 2019, 4.5% of the country's electricity was generated from nuclear power. With a total net capacity of 13.6 GWe, Canada has 19 operable nuclear reactors. In 2019, 15 percent of the electricity generated by nuclear power in the world.

With a total net capacity of 96.8 GWe, the USA has 95 operable nuclear reactors. Nuclear power provided 20% of the nation's electricity in 2019. With a total net capacity of 1.6 GWe, Argentina has three reactors. In 2019, the nation produced 6% of its nuclear power. There are two reactors in Brazil, with a combined 1.9 GWe net capacity.

In 2019, 3% of the nation's electricity was generated by nuclear power. There are seven operable nuclear reactors in Belgium, with a total net capacity of 5.9 GWe. In 2019, 48% of the electricity generated by nuclear power in the world.

With a total net capacity of 2.8 GWe, Finland has four operable nuclear reactors. Nuclear power provided 35% of the country's electricity in 2019. A fifth-1720 MWe reactor. France has 56 nuclear reactors which are operational, with a total net capacity of 61,4 GWe. Nuclear power provided 71% of the country's electricity in 2019. Germany, with a total net capacity of 8,0 GWe, continues to run six nuclear power reactors.

In 2019, 12.5% of the electricity in the country was produced by nuclear power. With a total net capacity of 45.5 GWe, China has 47 operable nuclear reactors. Nuclear power provided 5% of the country's electricity in 2019. India has 22 nuclear reactors which are operational, with a total net capacity of 6.2 GWe. In 2019, 3% of the nation's electricity was generated by nuclear power.

Japan has 33 nuclear reactors that are operational, with a total net capacity of 31,7 GWe. Just nine reactors were brought back online at the start of 2020, with a further 17 in the process of restarting the approval process following the Fukushima accident in 2011. In the past, 33% of the country's electricity came from nuclear power; in 2019, the figure was just 8% [4].

6. Is a nuclear energy renewable or nonrenewable source?

There are three main types of fossil fuels consider as nonrenewable energy sources: Coal, Oil and Natural Gas. They are nonrenewable energy sources because they exist in finite quantities. On the other hand, renewable energy means that they can naturally replenish themselves over time. Six main sources of renewable energy: Rain, Wind, Sunlight, Tides, Waves and Geothermal heat.

The researches with the nuclear energy as renewable energy source due to the following items:

Low-Carbon Emission: This is the main argument for nuclear energy being renewable. Nuclear power plants do not pollute the air or emit greenhouse gases [5].

It Is Replenishable: It takes more time than with the other sources, but in the end, they will appear again.

Fissile Material: Uranium supplies existing now can supply nuclear power only for approximately 1000 years.

While the against for the nuclear energy is consider the renewable source, due to:

Finite Uranium Deposit: Uranium deposit found on Earth is finite. Thus this resources one day will disappear, in addition to the following items:

- Nuclear power reactors give away harmful nuclear waste.
- Radioactive waste can be extremely toxic, causing burns and increasing the risk for cancers, blood diseases
- Storage of nuclear waste is very expensive.

Example the nuclear disasters that took place over the Chernobyl, Fukushima.

So, is nuclear energy renewable? There is no clear answer for that now. There are pertinent arguments on both sides of the debate [6, 7].

Author details

Nasser S. Awwad
Chemistry Department, Faculty of Science, King Khalid University, Saudi Arabia

*Address all correspondence to: nsawwad20@yahoo.com

References

[1] Uranium mining in Africa: past, present and future, Conference: International conference on the nuclear stations of Africa At: Johannesburg, South Africa, November 2015.

[2] "World Uranium Mining Production". London: World Nuclear Association. May 2020. Retrieved 2 September 2020.

[3] IAEA-TECDOC-1221 Safety of and regulations for nuclear fuel cycle facilities Report of a Technical Committee meeting held in Vienna, 8-12 May 2000.

[4] OECD International Energy Agency, World Energy Outlook 2020.

[5] Johnson K. "Is Nuclear Power Renewable Energy," Wall Street Journal, 21 May 2009.

[6] Cohen B.L. "Breeder Reactors: A Renewable Energy Source," Am. J. Phys. 51, 75 (1983).

[7] Kanter J. "Is Nuclear Power Renewable," New York Times, 3 Aug 09.

Chapter 2

Nuclear Power Plant or Solar Power Plant

Mostafa Esmaeili Shayan and Farzaneh Ghasemzadeh

Abstract

Both solar energy and nuclear energy face significant economic challenges. Sustainable energy costs have traditionally been greater than any of those associated with the growth of fossil fuel power generation, although the costs of renewable energy technologies (especially photovoltaic) have dropped. Furthermore, capital costs remain a big challenge in the nuclear generation. In many nations, the cost of building small nuclear power plants is quite large due to time, technology, and environmental and safety challenges for consumers. Such problems might not be as big for state-owned corporations or controlled industries for which utilities have quick access to cheap resources, and this partially explains why the interest for nuclear reactors in Asia is far greater than in the United States or Europe. Learning could help decrease costs for both types of technologies, but the track record for learning-by-doing in the nuclear sector is not good.

Keywords: solar energy, nuclear energy, renewable, power plants, technology

1. Introduction

The sun is a nuclear fusion reactor that contains gravity. It produces unimaginable quantities of energy. Solar energy is a very perfect source of power. It can be captured passively by solar panels or other collectors. When the collectors have been produced, there will be no carbon emissions or waste products [1]. There are no moving parts to hurt wildlife. There is no dependence on foreign entities. The energy is produced and delivered for free by the sun [2]. The uranium division begins progressing with the absorption of the smooth-moving neutron from the non-strong U-235 isotope. The obtained U-236 is split into Ba-139 and Kr-94 as well as three unfastened neutrons. The mass deficiency of approximately 20% of atomic mass units has also been converted into 210 MeV energy units [3, 4]. There were 447 nuclear fission power stations in service globally, 55 in construction and 111 in the design processes [5].

In the United States in 2018, 19.3% of the electricity supply was produced by 97 nuclear power plants. This amounts from zero percent to the other countries, for example, in New Zealand, and 71.7% in the European Union; the total global energy demand in 2018 was 10.3% [6].

With 11 new reactors under development, China has the most quickly expanding nuclear power program. Pakistan aims to construct three to four nuclear power stations by 2030 [7].

Several countries had nuclear installations in the past, but they still do not have nuclear plants in operation. Italy closed all the nuclear power stations between them by 1990, and, as a consequence of the referendums established by the Italians in 1987, nuclear power already has stopped [8]. A number of nations currently run nuclear power stations but are considering the process of nuclear technology. These countries are Belgium, Germany, Switzerland, and Spain [3]. Also according to the U.S. Energy Information Administration (EIA), solar power increased by 39% in the United States from 2014 to 2017 [4]. Starting at 10 GW and ending at 27, this growth trend for the field is very encouraging. In addition, carbon dioxide emissions have decreased by a few percent, the lowest since 1991 [5]. If it continues down this path, more study is likely to be carried out as a result of the growth in the market for efficient, cheap solar energy, in order to attempt to develop even more carbon-free or low-carbon fuels such as wind and nuclear power [6]. There are two big issues relating to nuclear plants: waste disposal and potential failure. Nuclear power plants produce dangerous wastes; for example, a 1-GW nuclear power plant can produce 300 kg of nuclear waste, with a half-life of almost 24,000 years, and cause environmental issues. The current methods for disposing of these kinds of waste are inadequate. The complete reprocessing of all radioactive waste and the chemical transformation of long fission products will be an ideal option. However, trends in this area have not progressed extensively [7]. The first and most critical problem is its disparity; the amount of solar energy that can be harvested depends widely on the time, location, season, weather, and several other factors. In order to improve this topic, engineers are exploring the development of new storage methods for large quantities of energy generated [5]. One of these storage techniques suitable for mountainous areas is pumped hydroelectric storage (PHES) that also uses excess energy generated during nonpeak hours of the day to pump water from a reservoir in a much high elevation. PHES is just one of the several potential storage methods used by many people, and it is so essential because it provides a clean, efficient use of solar energy when normally none is generated by replacing it with hydroelectricity [8]. Because of the good use and storage of solar energy, it becomes more difficult to determine whether to use solar energy or some other form of renewable energy for power companies and individuals. Despite the obvious cost of installing solar power, this is a higher investment opposed to the use of fossil fuels due to much lower maintenance and occasional overproduction of energy.

Solar energy is a key player in the sustainable power plan. In sunny places, many residents built panels on their roofs to support air-conditioning, heating, and other household needs and the panels were set up by themselves. Study in the collection and storage of solar energy should be a major effort worldwide [9]. But in less sunny areas, there are a few expensive homes which run 100% on solar power, using large battery banks to power them through the nights.

Solar energy has the capacity to boost everything we need; however our ability to turn the energy of the sun into electrical power and also to store energy is simply not fully developed. Energy storage in particular has proven to be challenging, as solar panels have a very irregular energy intake because it depends on season, climate conditions, time of day, and so on. The inability to use all solar power harvested efficiently is an issue that is likely to force even more development in the field to come soon after it has been resolved. The industry is full of possible innovations that have yet to be made and which can be recognized if time is taken to develop the innovative technology. Therefore, when looking at potential ways of storing the energy produced, PHES may not be the most cost-effective, but it is proven to be safe and can be added to some existing infrastructure at the same time as analysis seeks to make it more efficient.

2. Solar energy

Solar technology, i.e., renewable wind, offers a reliable and stable supply of solar energy during the year. As our natural resources are likely to decline in the years to come, it is necessary for the entire world to shift toward sustainable sources. Solar power is a reactive electromagnetic sunlight energy that can be used for a wide range of still-evolving applications, such as solar heating, photovoltaics, solar electricity, solar thermal processing, artificial molten, salt power plants, and photosynthesis.

Solar energy is a significant source of green energy, and its techniques are generally characterized as either passive solar energy or active solar energy based on whether solar energy is absorbed and transmitted or transformed into solar energy. Strong solar technologies involve the use of photovoltaic devices, concentrating solar power and solar water heaters to harvest electricity. Passive methods include the alignment of a system or building to the sun, the use of products with desirable light properties or thermal mass, and the construction of spaces that automatically disperse air.

2.1 Advantages of solar energy

The biggest advantage of solar energy is that it can be quickly installed by both home and business consumers, because it does not involve any major construction, such as in the case of wind and geothermal power stations. Solar energy not only benefits individual owners but also benefits the environment. **Figure 1** shows a simple model of a solar thermal system.

1. **No pollution**: Solar energy is a safe, nonpolluting, efficient, and green energy resource. This does not pollute the environment by producing poisonous pollutants, such as carbon dioxide, nitrogen oxide, and sulfur oxide. Solar energy does not need power and thus prevents the problems of shipping power or handling radioactive materials.

2. **Long-lasting solar cells**: Solar cells have two special features: first the lack of drive systems and second the minimal maintenance requirements. Then they have already got a longer life and they're more noticeable.

3. **Renewable source**: Solar energy is a sustainable energy source that can continue to generate power as long as there is light. While solar energy cannot be generated during the night and rainy days, it can be used again and again throughout the day. Solar energy from the sun is a steady and continuous source of electricity which can be used to harvest strength in remote areas.

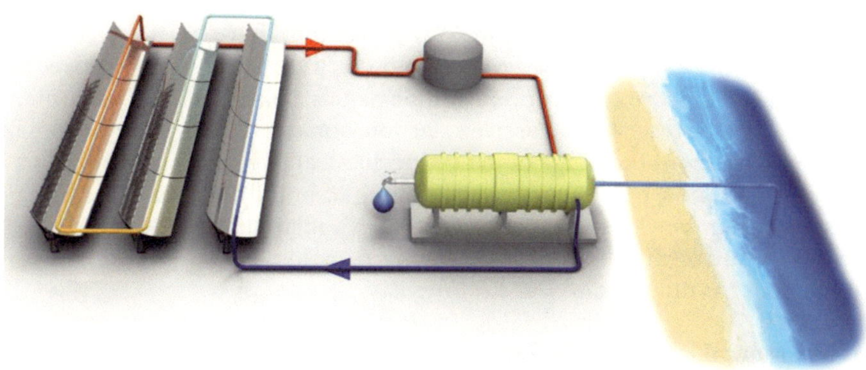

Figure 1.
Solar thermal system [10].

4. **Low maintenance**: Generally, solar cells do not need upkeep and operate for a long time. More solar panels can be installed from time to time if desired. While solar panels have an initial expense, there are no recurrent costs. The initial expense, which is paid once, may be recovered in the long run. Apart from this, solar panels do not create any noise and do not emit an unpleasant scent.

5. **Easy installation**: There is no need to install equipment such as cables, power supply, pipes, etc.; solar panels make solar tracking simpler. Unlike wind and geothermal energy harvesting systems that need land drilling equipment, solar panels do not need them and can be easily mounted on rooftops to insure that no additional infrastructure is needed, so residential home users can easily use this technology to supply electricity. In addition, they can be installed in a dispersed manner, meaning that no large-scale installations are required.

The technology of solar cells is developing, and as our nonrenewable supply decreases, it is necessary for the world to transition into renewable energy sources. There are, though, a range of issues that prohibit solar energy from being used more widely. Solar energy drawbacks are likely to be resolved as technology advances, and their use grows as people continue to realize the benefits of solar energy.

2.2 Disadvantages of solar energy

Solar energy can either be thermal or photovoltaic. The photovoltaic type is one of the most stable types of converting radiant energy into electrical energy. It really is suitable in many countries with adequate sunlight, such as Iran, and countries close to the equator, in terms of the quantity and availability of this technology. The energy source does not relate to someone and requires permission to use it. This feature has given rise to solar energy becoming special among renewable energy sources. Solar energy from ancient times is used by people using a magnifying glass to light the fire. Throughout this way, the sunlight was concentrated on dark wooden surfaces, and the fire became ignited. Also, solar photovoltaic (SPV) cells convert solar energy directly into DC electricity. This power source may be used to power solar clocks, calculators, or signals. These are also found in areas which are not linked to the power grid. **Figure 2** shows a concentrated solar power (CSP) plant. Solar heat energy (SHE) can be used to heat water or air, which requires ventilation of the room inside the house.

Solar energy can be broadly categorized as active or passive solar energy depending on how they are captured and utilized. For active solar power, specific solar heating equipment is used to transform solar power into thermal energy, but there is no specialized equipment for passive solar power [11]. Active solar requires the use of mechanical devices such as photovoltaic panels, solar trap fans, and solar thermal collectors or reservoirs. Passive solar solutions transform solar energy into thermal energy without the usage of active mechanical devices. It is primarily a method to use curtains, doors, plants, positioning of buildings, and other basic methods to catch or block the sun for usage. Passive solar heating is a smart way to save electricity and optimize its consumption. An example of passive solar heating is what happens to your car on a hot summer day.

2.3 Environmental impacts of solar power systems

Although solar energy is recognized to be one of the cleanest and most renewable sources of energy today, it also has several environmental impacts.

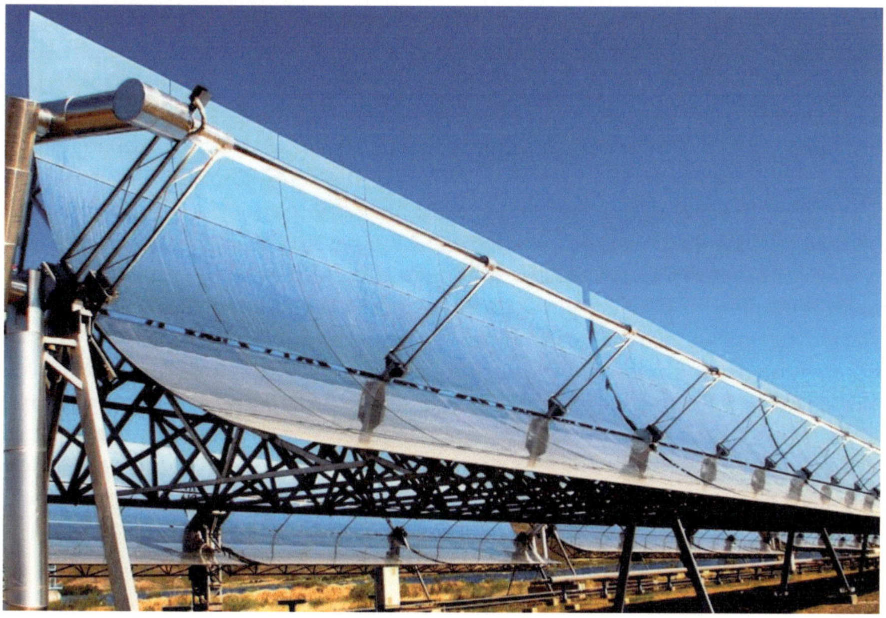

Figure 2.
Concentrated solar power (CSP) plant [10].

Solar energy uses photovoltaic panels to generate solar electricity. Nevertheless, the processing of photovoltaic cells to generate the energy includes silicon and to produce other waste products. Inappropriate handling of such materials can result in hazardous exposure to humans and the environment [12]. Installing solar power plants will entail a significant portion of land that may have an effect on established habitats. Solar energy does not pollute the air when converted to electricity by solar panels. It is found in abundance and does not help in global warming.

2.4 Solar energy's potential

Solar power is now expected to play a greater position in the future due to recent developments that will result in lower costs and better efficiency. In fact, the solar photovoltaic industry is preparing to supply half of all future US power generation by 2025. More and more architects understand the importance of active and passive solar power and know how to successfully integrate it into building designs. Solar hot water systems can compete economically with conventional systems in some areas. Shell has predicted that by 2040, 50% of the world's electricity supply would come from sustainable resources. Over recent years, the rate of generating photovoltaic cells has declined by 3% per year while policy subsidies have increased. While certain other information about solar energy is meaningless, this renders solar energy an even more efficient source of electricity. Solar energy is projected to be used by millions of households across the world in the next several years, as seen by developments in the United States and Japan. Aggressive financial incentives in Germany and Japan and China have made these countries global leaders in solar deployment for years [13].

A renewable resource that can be used to generate power is solar. The sun itself is a source of radiant, daylight, and other energy sources on Earth. Steam engines are a perfect illustration of radiant energy, by having sunrays magnified by mirrors guided to the turbine to heat water and produce steam, which in effect drives the turbine and causes steam to escape, and this pushes the piston. Calculators often work on solar power by storing light rays and transmitting energy to enable the calculator to function even though no light is present. Trevor Smith[1] notes that "solar rays can be used to fuel or cool houses, supply hot water and produce steam for turbines generating energy. Sunlight can be converted directly into energy by photovoltaics, a fast-growing branch of solar technology." This allows people to generate energy from renewable resources. James Bow notes that in 1977, 1 W of solar power costs $76.67. In 2014, the cost dropped to around $0.60. This suggests that modern solar power projects are far more economical, which means that renewable energy has come a long way and will continue to grow. One of the greatest declines in solar power is that, first, the sun is still growing and dropping, ensuring that the energy provided and processed is confined to the location of solar panels. Second, the batteries used to store electricity generated by the sun are expensive and produce a large amount of emissions. Third, in order to allow the best of the light, wide quantities of solar panels or mirrors need to be installed, which could be a function of restricted resources. The energy generated by the solar is a type of renewable energy used by today's society.

3. Nuclear energy

Nuclear power is the energy of an atom. Atoms are very tiny objects which make up a single body in the universe. There is enormous power in the links that connect the nucleus unchanged. Power is generated when the ties are disbanded. Nuclear energy may be used to create electricity, but it must be produced first. Nuclear power can be produced by both nuclear fusion and nuclear fission. In nuclear fission, atoms are separated into smaller atoms, which generate steam. Nuclear power stations have been used for electricity generation. Another method of generating nuclear energy is through nuclear fusion. The combination of atoms to each other and the creation of heavier atoms are established. When atoms are coupled, a lot of energy is released. These reactions occur together in the sun to generate thermal energy to radiation. Numerous studies are currently underway, although this technique has not yet been commercialized and it is not known if it is possible to generate electricity from this method. Uranium (U-235) is the most commonly produced nonrenewable material for nuclear fission. Plants use a particular type of U-235, as the atoms are readily isolated. During nuclear fission, the neutron hits and splits the uranium atom, releasing a large sum of energy in the form of heat and radiation. More neutrons are also released as the uranium atom is separated. Some neutrons proceed to hit other uranium atoms, and the process begins over and over again. It's a chain reaction, too. Although uranium is around 100 times more common than silver, U-235 is extremely scarce. Most of the US uranium is extracted in the western United States, but only 17 percent of the plutonium reactors is generated abroad. Uranium provided to US reactors in 2013 arrived from a number of nations, including Russia, Australia, and several other African countries. **Figure 3** displays the map of uranium mines in the world [14].

There are 648 nuclear power stations in the world. There are 61 nuclear power stations and 99 research facilities in the United States. Nuclear plants are found in 30 states, and 46 are situated east of the Mississippi River. After 1990, nuclear power has supplied around one-fifth of US electricity annually. Nuclear power provides as much electricity as all the fuel consumed in California, New York, and Texas together. Nuclear energy plants supply more than 20% of US energy. **Figure 4** shows the map of nuclear power stations in the world.

Figure 3.
Map of uranium mines in the world.

Figure 4.
Map of nuclear power stations in the world.

3.1 Nuclear power is the result of nuclear fission

Uranium fission occurs with the capture of the slow neutron by the non-isotope U-235. The resultant U-236 generates three free neutrons and separates into Kr-94 and Ba-139. The mass defect of roughly 0.2 atomic mass units is converted into 210 MeV energy units. $U = 1.66 \times 10^{-27}$ gk for the atomic mass unit, and eV equals 1.60×10^{-27}J, the radioactive energy unit.

Many power stations, like nuclear power plants, use heat to produce electricity. Power plants rely on steam from hot water to drive massive turbines, which then produce electricity. Because of using fossil fuels to produce electricity, nuclear power plants employ nuclear fission energy. The fission occurs in the nuclear power plant reactors. Nuclear reactors are devices which contain and regulate nuclear chain reactions while releasing heat at a regulated rate. The nucleus of the device, which includes nuclear fuel, is at the top of the plant. The uranium fuel is constructed of ceramic pellets. Each ceramic pellet contains at about the same amount of energy as 150 gallons of gasoline. Such energy-rich pellets are packaged in 12 foot wire fuel pipes. The array of fuel rods, sometimes hundreds of them, is called a burn unit.

The heat generated during the fission at the center of the reactor is used to boil water to steam, which turns the turbine blades. The energy can be generated while the rotor blades rotate. Afterwards, the steam is pumped back into the atmosphere in a different power plant system called a cooling tower. The product will be collected.

Nuclear power plants do not emit carbon dioxide emissions during operation compared to fossil fuel-fired power stations. Methods for the extraction and refining of uranium oxide and the processing of nuclear fuel, however, require a large amount of power. Nuclear power stations supply large quantities of metal and concrete which also require a substantial amount of energy to be produced. When fossil fuels are used for the production and refining of uranium oxide or for the installation of a nuclear power plant, the emissions generated by the burning of these fuels may be associated with the energy emitted by nuclear power plants. The main environmental concerns linked to nuclear power include the processing of toxic waste such as uranium mine tailings, expended reactor fuel, and other nuclear waste. These materials can stay radioactive and dangerous to human health for thousands of years. Animals are subject to strict laws governing their care, delivery, preservation, and treatment for the protection of human health and the environment. The US Nuclear Regulatory Commission (NRC) regulates the operations of nuclear power plants. Nuclear waste is classified as small and large rates of emissions. Radioactivity of these materials may range from just over natural background rates, including in mill tailings, to much higher amounts, such as spent nuclear fuel or sections of a nuclear plant. Radioactivity of toxic waste is decreased as time passes by a process called nuclear decay. The period of time taken to reduce the radioactivity of hazardous material to half of the original level is considered the contaminated half-life of the substance. Short-lived radioactive waste is also treated permanently prior to disposal in order to mitigate the future danger of contamination to staff handling and carrying waste, as well as to the amount of pollution at production sites.

Nuclear waste stored in tanks is very dangerous. These vessels are kept under special conditions in the water with safety shields until their half-life exceeds the standard of security. Various countries have specific laws on the processing of nuclear waste. The United States has set out strict rules on the storage and management of radioactive fuel and waste. Some nuclear power plant fuels can be stored in dry storage tanks. In this way, nuclear fuel tanks are stored in separate rooms with cement or steel air-conditioning devices.

Typically, once a nuclear reactor stops, it shifts. It involves the controlled extraction of the reactor and other devices that have been damaged from operation and the elimination of radioactivity to a degree that permits other uses of the site. The United States Nuclear Regulatory Commission (NRC) has stringent regulations regulating the decommissioning of nuclear power facilities, including the washing up of radioactively polluted reactor processes and equipment, including the disposal of atomic waste.

Uncontrolled nuclear reactions in a nuclear reactor will potentially contribute to extensive pollution of air and water. The probability of this occurring at nuclear power plants in the United States is known to be very low due to the complex and robust safeguards and multiple protection measures in effect at nuclear power plants, the preparation and expertise of reactor workers, the monitoring and service operations, and the legislative standards and oversight of the United States. A wide-field near nuclear power plant is controlled and supervised by trained security forces. Some of the reactors have containment vessels that are designed to withstand extreme weather events and earthquakes.

3.2 Advantages of nuclear energy

According to the laws of physics, energy is neither produced nor destroyed, but it can be converted from one kind to another, including the transfer of electrical energy into mechanical energy of electric motors. From the structure of the atom, much of its mass exists in a part called the core, and this mass contains protons with a positive electric field and neutrons with an ineffective or neutral electric field. Studies and experiments have indicated that neutrons weigh a lot more than protons. Nuclear energy is the energy generated by a nuclear explosion or a nuclear fusion under the specific conditions of the nucleus of an atom. A lot of energy can be released as nuclear fission or nuclear fusion happens. Once the heavy element, uranium, was exploded with neutrons, it was found that something special occurred instead of causing radioactivity as other materials. This cycle has been called fission. When nuclear fusion or nuclear fission happens as a product of neutron impacts, not only are two lighter elements produced and many radiations released, but more neutrons are generated, as can be seen in **Figure 5**. It is therefore obvious that concurrently released neutrons can start a chain reaction by acting on released light atoms, increasing the intensity of the reaction. This reaction may spread throughout uranium.

A lot of energy would be produced through the fission of the uranium-235 nucleus (see **Figure 5**). To consider the amount of this energy, it's enough to remember that this amount is around 60,000,000 times greater than when a carbon atom burns. During a nuclear fission reaction, the atom decomposes and releases a lot of kinetic energy into the environment. Obviously, kinetic energy is directly related to the generation of heat. The first reactors to generate a functional volume of electricity were installed in the Calder Hall in England. Atomic bombs may be produced of mere fissionable material. Of the two bombs dropped on Japan to end the World War 2, one contained plutonium and the other very highly enriched uranium-235.

3.3 Advantages of nuclear energy

1. ***Lower greenhouse gas emissions***: As recorded in 1998, the production of greenhouse gasses has been projected to have declined by almost half owing to the success of the usage of nuclear power. Nuclear processing has by far the lowest environmental impacts, because it does not release greenhouse gasses such as carbon dioxide, a fuel that is largely responsible for the greenhouse effect. Thanks to its application, there is no harmful impact on water, soil, or other environment, although certain greenhouse gasses are emitted when shipping fuel or harvesting uranium oil.

2. ***Powerful and efficient***: The other major benefit of having nuclear technology is that it is more effective and efficient than other potential forms of electricity. Technology advances have rendered it more competitive than most. That is one of the reasons that many nations are spending extensively in nuclear power. At least, a tiny part of the world's energy is flowing into it.

3. ***Reliable***: In comparison to conventional energy sources such as solar and wind, which involve sun or wind to generate electricity, nuclear energy may be generated from nuclear power plants even under extreme weather conditions. They also can provide 24/7 power and need to be shut down for maintenance purposes only.

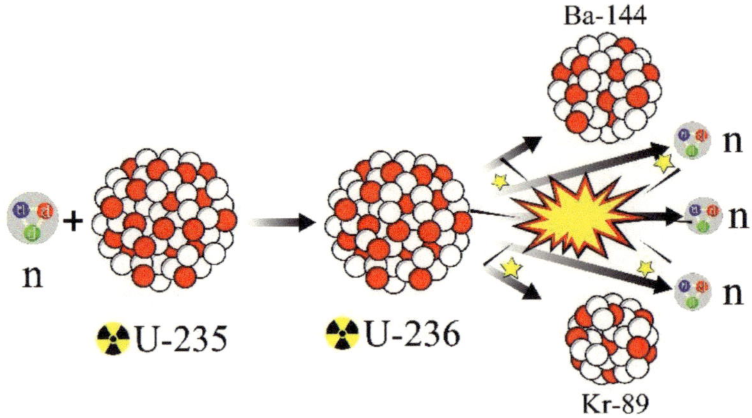

Figure 5.
Uranium-235 radioactive fission.

4. ***Cheap electricity***: Similar to traditional energy sources such as sun and wind, which require solar or wind power production, nuclear electricity may be produced from nuclear power plants even under severe weather conditions.

5. ***Low fuel cost***: The key factor behind the low cost of fuel is that it takes a limited amount of uranium to generate oil. When a nuclear reaction happens, it produces millions of times more hydrogen than normal energy sources.

6. ***Supply***: There are other economic benefits of building up nuclear power stations and utilizing renewable electricity instead of traditional oil. It's one of the nation's biggest producers of energy. The greatest part of it is that this electricity has a constant availability. This is readily accessible, has large supplies, and is projected to last about 100 years, whereas electricity, oil, and natural gas are small and are likely to disappear early.

7. ***Easy transportation***: Electrical power generation requires much fewer raw contents. This implies that just 28 g of U-235 produces as much energy as 100 metric tons of coal. As it is needed in limited amounts, the transport of fuel is much simpler than that of fossil fuels. Optimal use of natural resources in energy production is a rather careful approach for every country. This not only strengthens the socioeconomic climate but provides a precedent for other countries as well.

There is no question that nuclear technology has found its way into the future; however, like most electricity forms, it still suffers from certain significant disadvantages.

3.4 Disadvantages of nuclear energy

1. ***Radioactive waste***: Waste generated by nuclear reactors must be disposed of in a secure location because it is highly dangerous and may leak radiation if it is not properly treated. Any kind of pollution releases radiation from tens to hundreds of years. Collection of toxic waste has become a significant obstacle in the growth of nuclear programs. Nuclear waste includes radioisotopes with

lengthy half-lives. This ensures that the radioisotopes exist in one shape or another in the atmosphere. Such aggressive radicals pollute the sand or the sea. It's classified as mixed waste. Mixed waste induces toxic chemical reactions, which create harmful problems. Radioactive waste is normally covered beneath sand and is classified as proof, although the material is going to be used to produce atomic weapons or chemical bombs.

2. ***Nuclear accidents***: While too many modern measures have been placed in motion to insure that such a tragedy will not arise again as Chernobyl or, more recently, Fukushima, the danger associated with it remains fairly high. Just slight radiation exposure may have disastrous consequences. There are some symptoms that induce fatigue, vomiting, diarrhea, and exhaustion. Many operating in nuclear power plants that live in these areas are at risk of obtaining the toxic radiation on what they are consuming.

3. ***Nuclear radiation***: There are power reactors that are called breeders. They're making plutonium. It is an element that is not present in nature but is a fissionable product. It is a by-product of a chain reaction which, once added in nature, is very toxic. It is mainly used for the development of nuclear weapons. Very definitely, it's considered a dirty gun.

4. ***High cost***: Another realistic drawback to utilizing nuclear technology is that a lot of money is required to put up a nuclear power plant. This is not often feasible for developed nations to support such an expensive renewable energy source. Nuclear power plants usually take 5–10 years to build, because there are a variety of legal formalities to be done, so they are often protested by those residing nearby.

5. ***National risk***: Nuclear technology has provided humanity the ability to create more bombs than to generate anything that will render the planet a safer community to stay in. They ought to be more cautious and diligent when utilizing nuclear technology to prevent any big incidents of any sort. They are soft sites for terrorists and extremist groups. Health is a big concern here. A little weak protection will prove to be deadly and barbaric to humans and even to this world.

6. ***Impact on aquatic life***: Eutrophication is another consequence of nuclear waste. There are several workshops and conferences that take place every year to find a common answer. As of yet, there is no result. Studies claim the nuclear waste requires nearly 10,000 years to return to its original state.

7. ***Big impacts on health and medicine***: We still remember the horror that unfolded during the World War 2, after the atom bombs dropped on Nagasaki and Hiroshima. Still after five decades of mishap, children were born with defects. This is partly due to the nuclear influence. Will we have some treatments for that? The response is no.

8. ***Availability of fuel***: Given the abundance of fossil fuels in most countries around the planet, uranium deposits are so hazardous that they are only available in a few countries, as the map of accessibility to uranium resources depicts in **Figure 3**. Permissions from a variety of foreign bodies are needed before anyone would even conceive about constructing a nuclear power plant.

9. ***Nonrenewable***: Nuclear technology requires plutonium, which is a limited resource that has not been produced in many nations. Most countries depend on other countries for the continuous supply of this gasoline. It's extracted and shipped like any other tool. Supply should be secure as long as demand is accessible. Once all the nuclear reactors have been dismantled, they would not be of much benefit. Due to its dangerous effects and restricted availability, it cannot be identified as renewable.

Various nuclear energy projects are ongoing in both developed and emerging countries, such as India. Not to note, the benefits of nuclear technology are well ahead of the drawbacks of fossil fuels. That's why energy generation technology has been the most preferred technology.

4. Conclusions

By concatenating uranium extraction from seawater, manifestly safe breeding reactor technology, and borehole disposal of nuclear waste, a viable, planetary-scale nuclear energy network can be developed, i.e., another that is capable of supplying such an enormous quantity of energy at such a high degree of intensity that it can be relied on to sustain much—and possibly much—of the human society in virtually much possible scenarios of significant concern. For that way, nuclear technology is qualitatively distinct from other consumable technology options and must be assumed to be completely renewable in other respects. Throughout the immediate future, it is possible that the opportunity to build and demonstrate manifest protection for the latest generation of modern nuclear plants would be necessary to establish the basis for a prosperous future focused on nuclear technology.

Human civilization needs fossil energy because of its current facilities and its basic needs. This need and the high use of fossil fuels in the industrial, commercial, and residential sectors have contributed to major rapid climate change. The challenge of global warming is one of the massive problems confronting governments around the world. Earth heating may change the ecosystems and create many long-term problems. Greenhouse gasses like carbon dioxide are rising water levels in the oceans. Some droughts are in risk of extinction. These concerns are so significant that crisis analysts have described the modern century as a fuel for sustainability and protection of the planet.

Many countries have adopted official targets for the share of renewable energy in their grids, and others are considering them (**Figure 6**). Now that the governments of the world have a common issue, human beings will take collaborative action. Global organizations have been set up to manage this issue. The usage of renewable resources is one of the proposals created by global organizations to manage this crisis. Those alternative sources of energy include renewable energy and nuclear power. Countries must make decisions based on the long-term future to determine and improve the energy structures of the nation and calculate the various costs. At present, taking into account the cost factor, it is not possible to fulfill all energy demand from clean energy sources. But the good news is that this is possible with the cooperation of nuclear and renewable energy. Several countries have, in their perspective, made the energy demand share dependent on renewable and nuclear energy. Specific planners engage in predicting future projects and their costs. **Figure 6** constitutes some of the OECD-calculated costs. The key competition today is between solar and nuclear energy. The cost of using solar energy over active nuclear energy continues to be substantial and significant. They are also ideal for all levels of challenging electricity. Costs for involvement in the energy

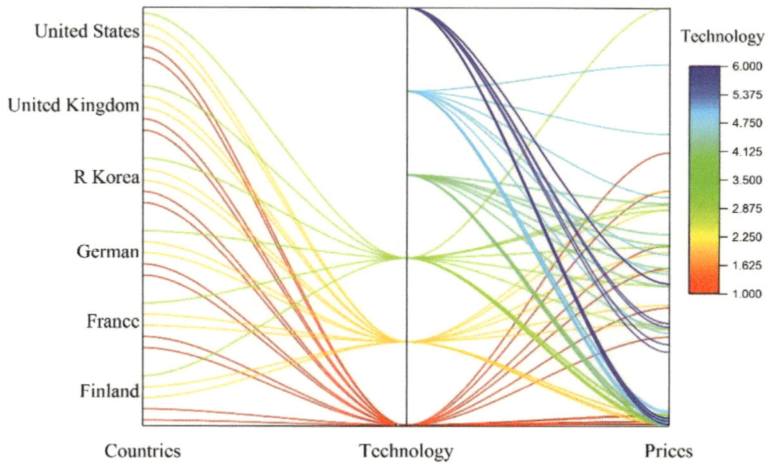

Figure 6.
Costs of combining nuclear technology with renewable energy.

market are assessed by the OECD per year. In **Figure 6**, the authors made the data comprehensible. Six countries pioneered the use of nuclear energy and renewable energy sources. **Figure 6** shows that the United States has been able to keep the cost of participating energy resources low, with the highest level of technology.

If the nuclear energy program is properly and sustainable way installed and the cost limit is eliminated, supplying electricity from nuclear energy resources will be reasonable and resolve these critical challenges for decades to come. It therefore seems necessary that we, as founders and citizens of a global society, begin to lay down the technological and structural foundations that will enable a viable, full-scale nuclear energy network to become operational in the immediate future while at the same time doing the same with regard to other realistic types of renewable energy supply on a scale [14].

One of the problems for nuclear power plants, as discussed earlier, is the difficulty of supplying 100 percent of electricity through these power plants. If we allow the setup and control share to be 10 percent and that share is given by solar energy, then the problem will be solved. However, if their share is assumed to be relatively large, then the cost of the system will increase, presenting another challenge.

These are rather heroic calculations given the paucity of sources, but they do indicate plausible effects. Increasing the penetration of renewable has small effect on backup costs since they tend to increase in direct proportion to the renewable capacity (MW) that needs backup and the increased capacity adds proportional MWh. However, balancing costs increase because more spinning reserve capacity is required at lower load factors. Since research is lowering the price of the solar-connected grid, the next problem is network costs. When renewable sources of energy such as photovoltaic systems manage to meet a district or village's full demands, then there will be a crisis. Power plants continue to use energy for spending networks indefinitely, and it is not clear how cost-effective these networks are. In this case, it would be illogical to establish and to develop a network [15]. (As a side note, backup and balancing are less costly in the United States because the dispatchable power is typically gas fired, which is less costly there.)

Another inference can be drawn from these results: the marginal cost of the system will generally increase with increased penetration of renewables, essentially due to their intermittentness and tendency toward remote locations. In addition to marginal system cost per MWH, there is another critical metric: marginal cost per ton of CO2 emissions reduced by increased deployment of renewables. After all, that is a primary policy driver for renewable targets.

The United States has set ambitious targets for renewable penetration: 33% by 2020, not including hydro. Further consideration (up to 51% in the legislative proposal) has been given for the future. The 33% level is thought to result in an implicit cost of $50/ton carbon reduction. Some energy companies have carried out important research on target utilization of 50% nuclear power and 50% solar energy. This research includes researching this topic in both scientific and economic terms. It was represented in **Figures 6** and 7. If the target is 50%, the lowest cost is $403, and the lowest cost is $340 for 40%. When energy storage technologies, such as nighttime high-altitude storage, are planned for solar energy, then the scenario will be more complex. This scenario shows that the size of large power plants can be utilized with good systems. For plants larger than 5000 mW, $636 per ton saves economic power [16]. **Figure** 7 shows the costs involved with combining nuclear and renewable energy.

The next challenge is solar and nuclear energy competition. Although solar power plants will fall in price each day, in most countries the price of renewable energy is still higher than in nuclear power plants. The cost of integrating and merging systems is also important. Currently, the value of building nuclear power plants in many countries is very high due to the companies concerns of moment, technology, sanctions, security, and safety hazards. It is possible to eliminate those limitations in solar energy. The same problems may not be as wide for state-owned companies or regulated markets that services have ready access to cheap capital, and that partly explains why Asia's enthusiasm for nuclear reactors is far stronger than it is in the United States or Europe. Researchers are working to reduce the costs of technology, but the nuclear industry is not strong, although that could improve small modular reactors if they can be developed in the process. According to **Figure** 7, given the right facilities, the United States has to pay the lowest costs for involvement in nuclear and solar energy. South Korea also has the right structure to take this scenario forward.

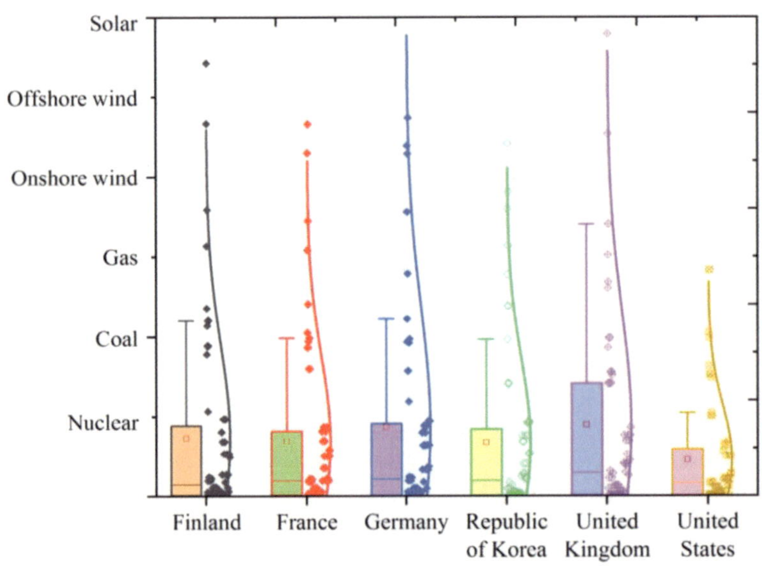

Figure 7.
The cost of getting a combination of nuclear energy and renewable energy paid by different countries.

Nuclear and renewable energies qualify for subsidies that vary from country to region. Some subsidies are direct, such as feed-in imports for renewable energy sources, while others shift the risks from utilities to customers.

The final guidelines would help to better compensate for nuclear and renewable costs and could help to reduce the costs of both:

1. Comparisons of nuclear and renewables costs should account for systems integration and differences in capacity factors.

2. In order to estimate nuclear costs, more attention should be given to the choice sensitivity of discount rate, as the discount rate drastically impacts the relative economic attractiveness of a nuclear project.

3. Findings on problems that may restrict the use of a nuclear reactor in "load-following" phase are important and should be given high priority.

Priority should be given to new reactor technologies like SMRs and regulatory reform in order to reduce nuclear capital costs. The final results of this section will include the following explanation. These explanations will help in choosing and policy making in the field of solar and nuclear energy. The outlook for these results is for the next 10 years. This outlook may change by changing conditions and creating critical conditions such as dramatically lower fossil fuel prices.

Nuclear power is dirty, dangerous, expensive, and not carbon-free and encourages nuclear proliferation. The nuclear power plant itself does not release toxic gasses such as CO2. Nevertheless, nuclear power leads to climate change; for any phase in the fuel chain used to produce electricity at the end of the day, a lot more energy is required, such as uranium extraction and uranium enrichment, which are highly energy-intensive methods. The life study of the whole fuel chain clearly indicates the relation to nuclear electricity to climate change. In a pioneering study [17], more than 100 studies have identified important but simple results, analyzing the life cycle of greenhouse gas emissions equivalent to greenhouse gasses produced at nuclear power plants around the world. The results show that if the life expectancy of a plant is equal to the greenhouse gas emission equivalent to that energy production, then the emission equals 1.4 g of carbon dioxide per kilowatt hour (gCO_2e/kWh) up to 288 gCO_2e/kWh is variable. The mean greenhouse gas emissions equate to 66 gCO_2e/kWh.

As a first conclusion, the extensive use of solar energy services for at least the next decade may be out of the issue. Photovoltaic and solar thermal systems, especially large thermal, wind, and biomass systems, will enter and expand energy networks quickly. Other renewable energy systems will be developed and priced to reduce consumption, such as biogas (wastewater, landfills, and livestock), geothermal, and possibly wave and tidal energy. This growth will be high in the next 10 years, but market with conventional systems will still take time [18]. Nuclear power is also an option when contemplating a transition from the dirtiest of fossil fuels, and thus nuclear power should be debated together with renewables. Nuclear time for building, risk, waste, and, in particular, costs must be tracked, because nuclear costs are increasing when solar energy costs are dropping. Small- and large-scale renewable energy projects and emerging storage systems are being increasingly developed by communities and nations. Also China, probably the most ambitious nation in terms of nuclear power, is introducing more wind and solar power relative to nuclear power—and not just nameplate capacity—which is actually produced. Last year alone, China installed 20.72 GW of wind (4.8 GW of production while its power factor is just 23%) and 28 GW of renewable energy

(10.6 GW of production), with about 90% of its solar installations coming from utilities. In the same year, more than five nuclear plants (5.7 GW output) were added to the existing wind and solar power. China is only one example of how wind and solar power can be installed quickly while producing more electricity. At the period (and if) China finishes its 28 nuclear power plants (many are still behind schedule), with an estimated potential of 34 GW, further wind and solar power would be installed around the same timeframe—again, taking into account efficiency factors [19].

For the coming 10 years, here in the United States, the five US nuclear power facilities are 2 years behind track and have a budget of billions of dollars. Once live, they will produce 5.1 GW while renewables would produce a rather modest 131 GW.

The other two factors are systems for the energy, safety and security systems. In a nuclear power plant, when things go awry, it can be really bad because of accidents, threats, or critical situations that happen. It should be noted that the smallest incident in a nuclear power plant can often incapacitate or destroy a city or a country. Is it likely? Who knows for sure? Could you foresee the next earthquake in Southern California or somewhere else in the United States or Japan or the rest of the world? What about the next wave washing down a coastline? How about the next cyber threat or the Middle East militant organization? Compare a tragedy for a nuclear power plant against a solar power plant. When you ask me why I'm against constructing new reactors, it's about economy, health and protection, and the reality that we can expand on current hydro and nuclear power facilities with all the renewables—and we can do it quicker.

Author details

Mostafa Esmaeili Shayan[1*] and Farzaneh Ghasemzadeh[2]

1 Tarbiat Modares University, Tehran, Iran

2 Iran University of Science and Technology, Tehran, Iran

*Address all correspondence to: mostafa.esmaeili@modares.ac.ir

References

[1] Esmaeili Shayan M, Najafi G, Ahmad BA. Power quality in flexible photovoltaic system on curved surfaces. Journal of Energy Planning and Policy Research. 2017;**3**:105-136

[2] Esmaeili MS, Najafi G. Energy-economic optimization of thin layer photovoltaic on domes and cylindrical towers. International Journal of Smart Grid. 2019;**3**:84-91

[3] Ogland-Hand JD, Bielicki JM, Wang Y, et al. The value of bulk energy storage for reducing CO_2 emissions and water requirements from regional electricity systems. Energy Conversion and Management. 2019;**181**:674-685

[4] Norman C. Nuclear Safety. Butterworth Heinemann. Elsevier Ltd.; 1974

[5] Alonso G. Desalination in Nuclear Power Plants. UK: Woodhead Publishing; 2020

[6] Yu Q, Zhang T, Peng X, et al. Cryogenic energy storage and its integration with nuclear power generation for load shift. In: Storage and Hybridization of Nuclear Energy: Techno-economic Integration of Renewable and Nuclear Energy. Elsevier Ltd.; 2018. pp. 249-273

[7] Ojovan MI, Lee WE, Kalmykov SN. An Introduction to Nuclear Waste Immobilisation. Elsevier Ltd.; 2013. pp. 1-362

[8] Biberian J-P. Cold Fusion Advances in Condensed Matter Nuclear Science. Elsevier Ltd.; 2020

[9] Murray RL, Holbert KE. Nuclear Energy: An Introduction to the Concepts, Systems, and Applications of Nuclear Processes. Elsevier Ltd.; 2019

[10] Esmaeili Shayan M, Najafi G, Gorjian S. Design Principles and Applications of Solar Power Systems (In Persian). 1st ed. Tehran: ACECR Publication-Amirkabir University of Technology Branch; 2020

[11] Aleixandre-Tudó JL, Castelló-Cogollos L, Aleixandre JL, et al. Renewable energies: Worldwide trends in research, funding and international collaboration. Renewable Energy. 2019;**139**:268-278

[12] Kerlin TW, Upadhyaya BR. Dynamics and Control of Nuclear Reactors. Elsevier Ltd.; 2019. pp. 95-125

[13] Fundamentals of Thermal and Nuclear Power Generation. Elsevier Ltd.; 2020

[14] Sanders MC, Sanders CE. Nuclear waste management strategies: An international perspective. 2019

[15] Zhong RZ, Cheng L, Wang YQ, et al. Effects of Anthelmintic Treatment on Ewe Feed Intake, Digestion, Milk Production and Lamb Growth. Singapore: Springer Verlag; 2017

[16] Suman S. Hybrid nuclear-renewable energy systems: A review. Journal of Cleaner Production. 2018;**181**:166-177

[17] Lima MA, Mendes LFR, Mothé GA, et al. Renewable energy in reducing greenhouse gas emissions: Reaching the goals of the Paris agreement in Brazil. Environmental Development. 2020;**33**:105-115

[18] Azadbakht M, Esmaeilzadeh E, Esmaeili-Shayan M. Energy consumption during impact cutting of canola stalk as a function of moisture content and cutting height. Journal of the Saudi Society of Agricultural Sciences. 2015;**14**:147-152

[19] Review and outlook of world energy development. Non-Fossil Energy Development in China. 2019:1-36

Chapter 3

Nuclear Fuel Transmutation

Akbar Abbasi

Abstract

Nuclear power plants to generates electric energy used nuclear fuel such as Uranium Oxide (UOX). A typical VVER–1000 reactor uses about 20–25 tons of spent fuel per year. The fuel transmutation of UOX fuel was evaluated by VISTA computer code. In this estimation the front end and back end components of fuel cycle was calculated. The front end of the cycle parameter are FF requirements, enrichment value requirements, depleted uranium amount, conversion requirements and natural uranium requirements. The back-end component is Spent Fuel (SF), Actinide Inventory (AI) and Fission Product (FP) radioisotopes.

Keywords: nuclear power plant, nuclear fuel, front end, back end, actinide inventory

1. Introduction

VVER –1000 (Water-Water Energetic Reactor-1000) is a type of pressurized water reactor with 1000 MW thermal power planned to generate a 330 MWe [1]. Production actinide consequently of using nuclear power reactors as electric energy source. Actinide Inventory (AI) elements cumulative in spent fuel (SF) and are a part of spent fuel that useable as MO_X fuel in nuclear power reactors. Recently, some researchers have been studied the actinide inventory in spent fuel of nuclear power reactors [2–4].

VISTA computer code is available for the calculation of nuclide inventories in spent fuel. The neutron transmutation (fission) of the long-lived actinide isotopes in SF with decay times on the order of millennia into fission products with decay times of a few hundred years would profoundly impact the problem of storing SF that confronts the expansion of nuclear power. For the actinides, the creation comprises of neutron catch or decay of a forerunner nuclide. Evacuation may comprise of neutron-actuated or unconstrained fission; neutron catch and radioactive decay [5].

The estimation of the response rates requires nuclide fixation and cross-area information, the neutron transition level and vitality range in the fuel. As the energy spectrum in the fuel is subject to the grid structure and arrangement, such counts include rehashed iterative answers for the range and cross-section. The degree to which this is completed relies upon the precision expected of the last arrangement. After every burnup span, the combined range is utilized to get the neutron cross-segments which are accordingly utilized for the count of the nuclide response rates. The focuses to be considered in making an assessment of the accessible strategies are:

- data of nuclide cross-section
- energy spectrum evaluation
- neutron flux level calculation during the irradiation
- burnup equations numerical solution.

The treatment of these amounts in the few elective codes has been analyzed [6].

2. Nuclear fuel cycle

Nuclear fuel cycle definition is the set of cycles to utilize nuclear materials and to restore it to conclusive state. The fuel cycle begins with the mining of unused atomic materials from nature and closures with the protected removal of utilized nuclear materials in nature. **Figure 1** shows the nuclear fuel cycle diagram by indicating main processes in a recycle mode.

The first step is mining in a nuclear fuel cycle. After this step the next step is milling prosses. The feed for mining and processing measure is U metal and the item is U_3O_8 concentrate, which is generally called yellowcake because of its shading and shape [7]. The third step is change term that alludes to the way toward purging the U concentrate and changing over it to the synthetic structure required for the following phase of the nuclear fuel cycle.

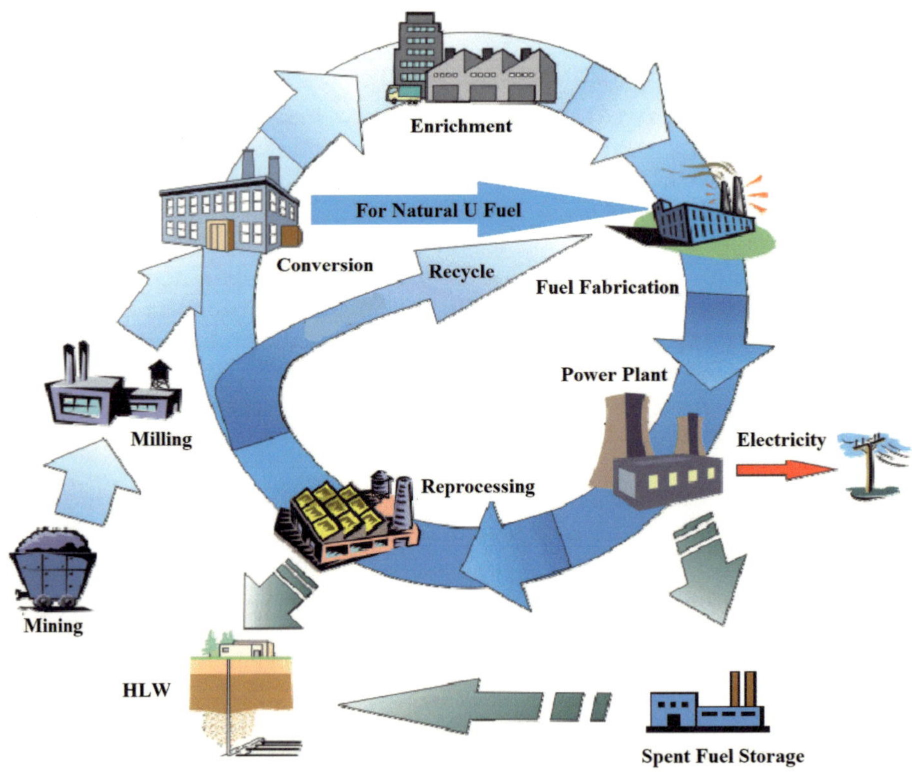

Figure 1.
The nuclear fuel cycle diagram.

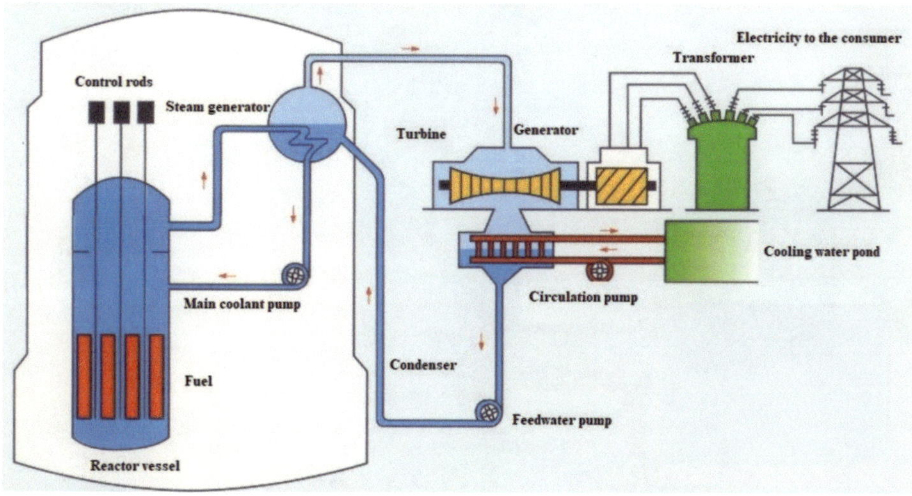

Figure 2.
Main components of a light water reactors (LWR) [8].

In this stage U element can be produced in three forms of metal, oxide (UO_2 or UO_3) and uranium hexafluoride (UF_6). UF_6 is the overwhelming item at this phase of the nuclear fuel cycle since it is handily changed over to gas for the advancement stage, as utilized on the planet's most regular reactor type. (LWRs) (see **Figure 2**).

The next process after conversion is enrichment step. In general, there are two industrially accessible advancement innovations: vaporous dispersion and rotator. The two strategies depend on the slight mass contrast somewhere in the range of ^{235}U and ^{238}U. Along these lines, the improvement is characterized as the way toward expanding the measure of ^{235}U contained in a unit amount of uranium. The feed for this stage is regular UF_6 and the item is enhanced UF_6. The other yield of the cycle is the uranium which has lower ^{235}U substance than the regular uranium. It is known as enhancement tail or exhausted uranium. Fuel fabrication is another term that the enrichment fuel was made as pellets. Fuel pellets are loaded into tubes of zircaloy or stainless steel, which are sealed at both ends. These fuel rods are spaced in fixed parallel arrays to form the reactor fuel assemblies (see **Figure 3**).

The whole process is referred as fuel fabrication. The reactor unit itself is irradiator for nuclear fuel. It burns the fuel, produces energy and spent fuel. The feed for reactor is new fuel containing U or U/Pu, if there should arise an occurrence of blended oxide (MOX) fuel, for existing atomic fuel cycle alternatives. The item is the spent fuel comprising of recently created nuclides, for example, splitting items (I. Cs, Sr, ...) minor actinides (Np, Am, Cm) and Pu just as the uranium. The greatest aspect of the spent fuel is still U (over 95% for the most reactor types). Reprocessing process is based on chemical and physical processes to separate the required material from spent nuclear fuel. The feed of this process is spent fuel and the products are reusable material and high-level wastes (HLW) [6].

The other unit of nuclear cycle fuel is spent fuel storage, which could be put away briefly for some time later or could be put away uncertainly. Spent fuel could be put away in pools (wet sort, briefly) or in storehouses (dry sort). Likewise, the loss from fuel manufacture and reprocessing offices are delegated HLW and requires cautious treating. HLW is put away in uncommon storerooms after legitimate treatment.

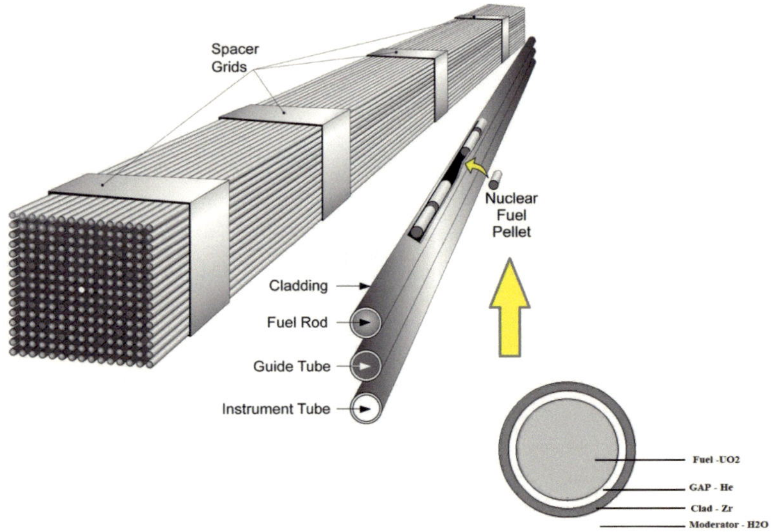

Figure 3.
The fuel fabrication [9].

3. The composition of transuranic in the spent fuel of VVER reactor

The following nuclides have been studied and the transmutation chain which is given in **Figure 4**. These radionuclides are: ^{235}U, ^{236}U, ^{238}U, ^{238}Pu, ^{239}Pu, ^{240}Pu, ^{241}Pu, ^{242}Pu, ^{237}Np, ^{241}Am, ^{242m}Am, ^{243}Am, ^{242}Cm and^{244}Cm.

The actinide transmutations to each chine are calculated by [10]:

$$\frac{dN_i}{dt} = -\sum_{i \neq j}\left[\lambda_{ji}^{d} + \sigma_{ji}^{tr}\varphi\right]N_i + \sum_{j \neq i}\left[\lambda_{ij}^{d} + \sigma_{ij}^{tr}\varphi\right]N_j \tag{1}$$

where N_i is atomic content of ith –isotope; λ_{ji}^{d} is decay constant, (1/s); σ_{ji}^{tr} transmutation cross section from isotope i to isotope j, (barn) and φ is average neutron flux, (n/s·cm^2).

If the neutron flux and cross sections are constant on a time interval, the equation has a simple analytical solution.

An example to solve the transmutation chain starting from ^{238}U up to ^{240}Pu is shown below, using Bateman's Equation.

$$^{238}\mathrm{U} \rightarrow {}^{239}\,Pu \rightarrow {}^{240}\,Pu \tag{2}$$

$$\mathrm{AF_1....AF_2....AF_3} \tag{3}$$

$$AF_1 = AF_1(initial).e\left(-\sigma_{t1}.\Phi.T.10^{-24}\right) \tag{4}$$

$$AF_2 = AF_1(initial).\left[\left(\frac{\sigma_{c1}}{\sigma_{t2}-\sigma_{t1}}\right).e^{-\sigma_{t1}.\Phi.T.10^{-24}} + \left(\frac{\sigma_{c1}}{\sigma_{t1}-\sigma_{t2}}\right).e^{e^{-\sigma_{t2}.\Phi.T.10^{-24}}}\right] \tag{5}$$

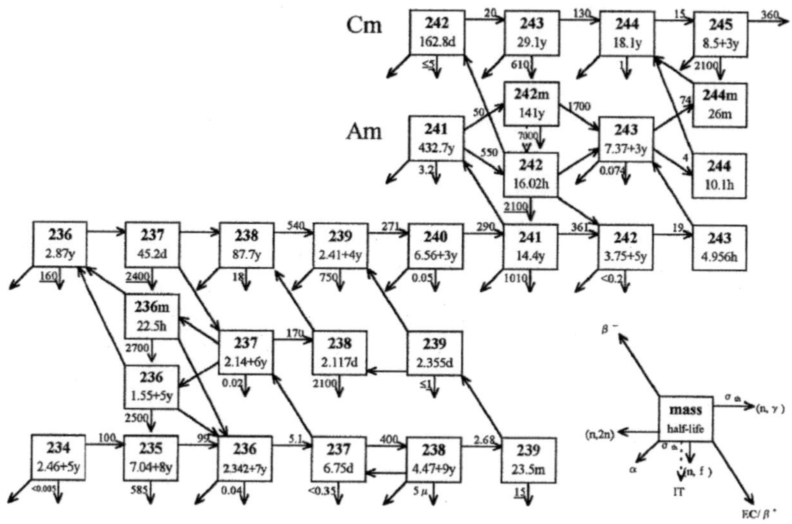

Figure 4.
The actinide transmutation chains [6].

$$AF_3 = AF_1\left(initial\right).\left[\left(\frac{\sigma_{c1}.\sigma_{c2}}{\left(\sigma_{t2}-\sigma_{t1}\right).\left(\sigma_{t3}-\sigma_{t1}\right)}\right).e^{-\sigma_{t1}.\Phi.T.10^{-24}} + \left(\frac{\sigma_{c1}.\sigma_{c2}}{\left(\sigma_{t3}-\sigma_{t2}\right).\left(\sigma_{t1}-\sigma_{t2}\right)}\right).e^{e^{-\sigma_{t2}.\Phi.T.10^{-24}}} + \left(\frac{\sigma_{c1}.\sigma_{c2}}{\left(\sigma_{t1}-\sigma_{t3}\right).\left(\sigma_{t2}-\sigma_{t3}\right)}\right).e^{e^{-\sigma_{t3}.\Phi.T.10^{-24}}}\right] \quad (6)$$

where.
AF_i = Isotope(i) atomic content in the chain
σ_c = Cross-section of capture (barns)
σ_f = Cross-section fission (barns)
$\sigma_{n,2n}$ = Cross-section of (n,2n) (barns)
σ_{ex} = Cross-section of excited (barns)
σ_t = Cross-section totally (barns)

$$\sigma_{decay} = \frac{0.693}{T_{\frac{1}{2}}.365.24.3600.10^{24}.\Phi} \quad (7)$$

$$\sigma_t = \sigma_c + \sigma_f + \sigma_{ex} + \sigma_{decay} \quad (8)$$

$T_{1/2}$ = Half-life (years)
Φ = Neutron average flux (n/cm/cm/sec). (the energy range of 0 to 10 MeV total flux)
T = Time of irradiation (sec)

Radiation	Mass (u)	Charge	Range (air)	Range (tissue)
α	4	+2	~3 cm	~40 μm
β	$\frac{1}{1840}$	-1 or + 1	~300 cm	~5000 μm
X or gamma emission	0	0	Very large	Through body
Fast neutron (n)	1	0	Very large	Through body
Thermal neutron (n)	1	0	Very large	~15 cm

Table 1.
Properties of nuclear emission.

$$T = \frac{1000.E_d}{3600.KWKG.24} \quad (9)$$

E_d = Burnup discharge (GW·d/t)
KWKG = Specific power (MW/tonne)

The condition solver initially computes the isotopic piece in nuclear division. The acquired nuclear portions at that point are changed over to the weight divisions [6].

Nowadays, it is estimated that >2000 t of actinides has been accumulated as nuclear waste, most of which are plutonium isotopes. **Table 1** shows the composition of transuranic elements in the fresh and spent fuel of a VVER after recycling process [10]. The most significant commitment to the drawn-out radiation peril originates from ^{239}Pu (t½ = 24,110 a), from other Pu isotopes, and from other actinides, i.e., ^{237}Np (t½ = 2.1 × 106 a), ^{241}Am (t½ = 432 a), ^{243}Am (t½ = 7370 a) and ^{245}Cm (t½ = 8500 a) [11]. Pu and MA represent only 1.5% of the waste volume. Nonetheless, their radio toxicity becomes dominant after around 300 years and remains extensively high for a huge number of years, a period too long to even

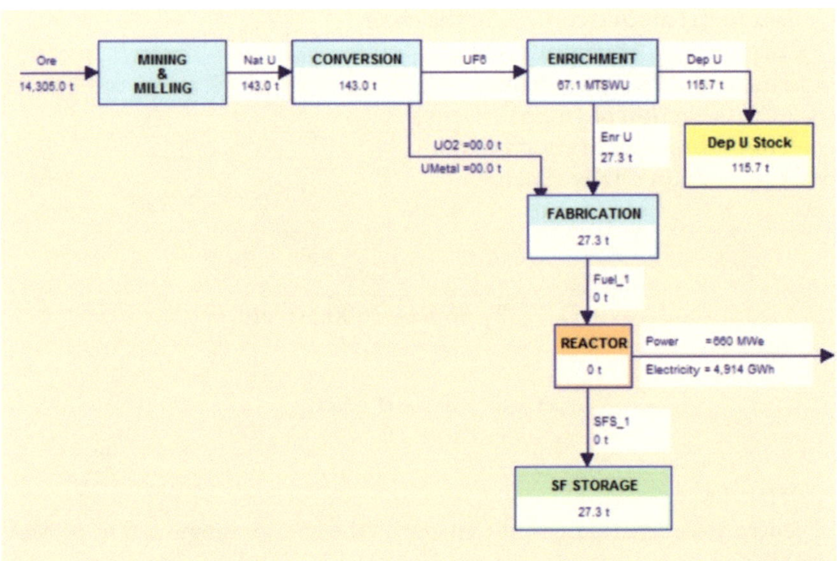

Figure 5.
The flowchart of nuclear material amounts calculated by VISTA.

consider guaranteeing a sheltered disengagement from nature by methods for building obstructions [12]. Besides, actinides present criticality and multiplication concerns. The fission cross-section of numerous actinides is portrayed by edges of a couple of 100 keV. Hence, they do not undergo fission in thermal reactors, rather reduce reactor critically as thermal neutron absorbers. However, they have significantly high fission cross-sections at high neutron energies [13].

The amount of nuclear materials for a VVER-1000 reactor was calculated and shown as diagram in **Figure 5**.

For VVER-1000 reactor, the fresh fuel, actinide elements and fission product values in spent fuel was calculated by VISTA simulation code.

The total amount of FF is 23.792 t/year with 22.915 t/year of ^{238}U and 0.877 t/year of ^{235}U. The grade of enrichment is 3.6% on average. The actinide martials content in SF of calculated by VISTA are ^{235}U (0.232123 t/year), ^{236}U (0.107850 t/year), ^{238}U (22.177277 t/year), ^{238}Pu(0.004352 t/year), ^{239}Pu(0.156181 t/year), ^{240}Pu(0.047959 t/year), ^{241}Pu(0.049525 t/year), ^{242}Pu(0.017008 t/year), ^{241}Am(0.001297 t/year), ^{237}Np(0.001239 t/year), ^{242m}Am(0.000019 t/year), ^{243}Am(0.003554 t/year), ^{242}Cm(0.000463 t/year) and ^{244}Cm(0.001142 t/year) radioelements. The values of above radioelements except ^{235}U and ^{238}U isotopes were compared in **Figure 6**.

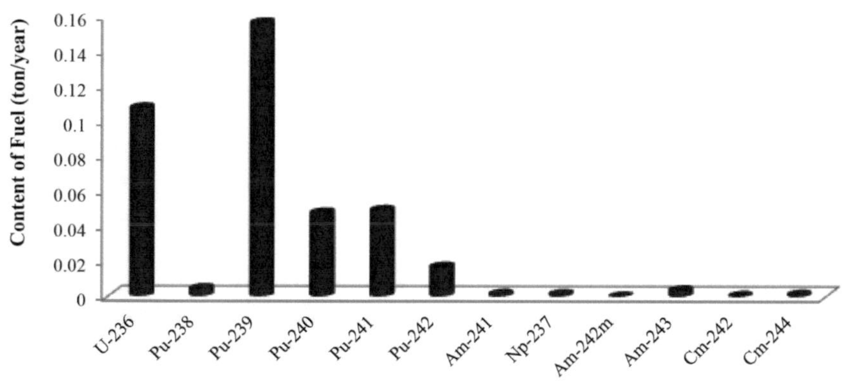

Figure 6.
The actinide elements content in spent fuel of the VVER-1000 reactor calculated by VISTA.

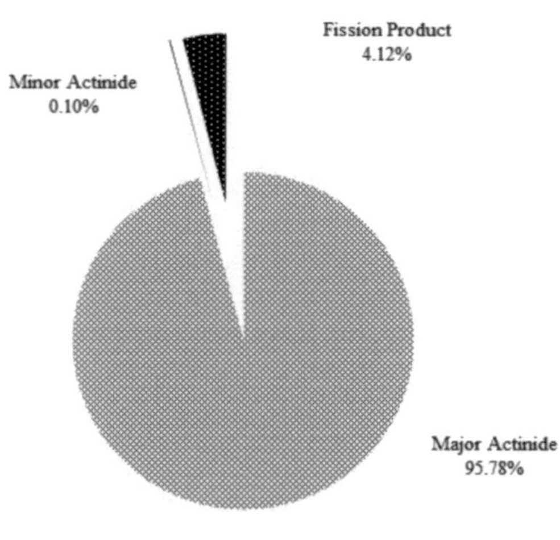

Figure 7.
Discharged UO_X spent fuel content in VVER-1000 reactor.

Also, the content of discharged UOX burned fuel in VVER-1000 nuclear power plant is presented in **Figure 7**.

4. Radiation and protection of nuclear fuel cycle

There are two type radiation sources naturally occurring radioactive materials (NORM) and technologically enhanced naturally occurring radioactive materials (TENORM) consist of materials in nuclear industry. The NORM radionuclides like ^{232}Th' ^{238}U, and ^{40}K that occur mostly in minerals such present all over the Earth's crust in varying quantities depending on the ambient geological end geochemical properties of local. NORM radioactive are present in soil [14–19], water [20–23] and building materials [24–30]. The TENORM materials is upset or changed from regular settings or present in a mechanically improved state due to past or introduce human exercises and practices, which may bring about a relative increment in radionuclide fixations, radiation presentations and dangers to people in general, and danger to the open condition above foundation radiation levels.

The properties and ranges of the various nuclear radiations are summarized in **Table 1**. The ranges are only approximate since they depend on the energy of the radiation [31].

The alpha particle has mass higher than beta particle, so these partials travels relatively slowly into matter. Alpha particle interaction is a high likelihood of with iotas along its way and will surrender a portion of its vitality during every one of these cooperation's. As an outcome, α particles lose their vitality quickly and travel without a doubt, extremely short separations in thick media.

Beta particles are a lot of littler than particles and travel a lot quicker. They consequently go through less associations per unit length of track and surrender their vitality more gradually than α particles. This implies β particles travel further in thick media than α particles.

Gamma radiation loses its vitality mostly by interfacing with nuclear electrons. It ventures enormous separations even in thick media and is hard to ingest totally.

Neutrons surrender their vitality through an assortment of collaborations, the general significance of which are reliant on the neutron vitality. Therefore, it is regular practice to separate neutrons into in any event three vitality gatherings: quick, moderate and warm. Neutrons are infiltrating and will travel enormous separations even in thick media.

An office ought to have set up a radiation assurance program that is satisfactory to secure the radiological wellbeing and wellbeing of laborers and the general population and guarantee that the presentations are ALARA. To achieve this, offices assess and describe the radiological hazard and regularly give adequate hearty controls to limit this danger. Potential mishap arrangements are considered in evaluating the ampleness of the controls, which expect to limit radiological danger and sullying.

The fuel cycle office radiation assurance rehearses incorporate [32]:

- A viable reported program to guarantee that word related radiological introductions are ALARA;

- An association with sufficient capability prerequisites for the radiation insurance work force;

- Approved composed techniques for directing exercises including radioactive materials;

- Radiation protection preparing for all faculty who approach limited zones;
- A program to control airborne convergence of radioactive material with building controls and respiratory insurance;
- A radiation overview and checking program that incorporates prerequisites for control of radioactive sullying inside the office and observing of outside and inward radiation presentations;
- Other projects to look after records, to report radiation introductions to the managing authority, and to restore an adequate in-plant radiological condition in case of an occurrence.

The execution of such projects with respect to coordinate radiation is currently made a lot simpler with the utilization of individual electronic dosimeters of Visa size that can immediately alarm the holder when momentary or cumulated portion reach modified edges, that keep in memory the historical backdrop of presentation and whose information can be downloaded to PCs, for instance each time the administrator enters or leaves the controlled zone, so these information can be naturally recorded and investigated. In this manner, point by point presentation previsions can be checked versus real introductions, permitting improvement of both working techniques and previsions. The improvement of mechanized screens that permit the perception of portion rates is likewise an incredible asset for radiation protection.

5. Conclusions

The content of this chapter is overall reviewing the nuclear fuel transmutation discussion. For this purpose, the nuclear fuel cycle of UOx type fuel was presented. In the next section the composition of transuranic in the spent fuel of VVER reactor was survived. Also, the amount of minor actinide and fission product in a VVER-1000 reactor was calculated and finally, the radiation protection principles of nuclear fuel cycle were presented and discussed.

Acknowledgements

The author would like to appreciate from to H. Tulsidas, Division of Nuclear Fuel Cycle and Waste Technology of IAEA.

Conflict of interest

The authors declare no conflict of interest.

Author details

Akbar Abbasi
Faculty of Engineering, University of Kyrenia, Kyrenia, Mersin, Turkey

*Address all correspondence to: akbar.abbasi@emu.edu.tr

References

[1] Mirekhtiary, Seyedeh Fatemeh, and Akbar Abbasi. "Uranium oxide fuel cycle analysis in VVER-1000 with VISTA simulation code." In AIP Conference Proceedings, vol. 1935, no. 1, p. 100005. AIP Publishing LLC, 2018. DOI: 10.1063/1.5025993

[2] Coates, David J., Benjamin A. Lindley, and Geoffrey T. Parks. Actinide breeding and reactivity variation in a thermal spectrum ADSR–Part 1: Development of a lumped thermal reactor model. Annals of Nuclear Energy 38, no. 10 (2011): 2120-2131. DOI:10.1016/j.anucene.2011.06.028

[3] Coates, David J., and Geoffrey T. Parks. Actinide evolution and equilibrium in fast thorium reactors. Annals of Nuclear Energy 37, no. 8 2010: 1076-1088. DOI: 10.1016/j.anucene.2010.04.004

[4] Zheng, Meiyin, Wenxi Tian, Dalin Zhang, Suizheng Qiu, and Guanghui Su. Minor actinide transmutation in a board type sodium cooled breed and burn reactor core. Annals of Nuclear Energy 81 (2015): 41-49. DOI:10.1016/j.anucene.2015.03.024

[5] Hu, Wenchao, Bin Liu, Xiaoping Ouyang, Jing Tu, Fang Liu, Liming Huang, Juan Fu, and Haiyan Meng. "Minor actinide transmutation on PWR burnable poison rods." Annals of Nuclear Energy 77 (2015): 74-82. DOI: 10.1016/j.anucene.2014.10.036

[6] INTERNATIONAL ATOMIC ENERGY AGENCY, Nuclear Fuel Cycle Simulation System: Improvements and Applications, IAEA-TECDOC-1864, IAEA, Vienna 2019.

[7] Ceyhan, M. "Modelling for nuclear material flows in the nuclear fuel cycle." In Fissile material management strategies for sustainable nuclear energy. Proceedings of a technical meeting. 2007.

[8] http://www.nucleartourist.com/type/vver.htm

[9] https://www.world-nuclear.org/our-association/publications/technical-positions/how-is-used-nuclear-fuel-managed.aspx)

[10] Abbasi, Akbar. "Analysis of uranium oxide fuel transmutation in VVER-1000 reactor using VISTA and WIMS-D4 codes." Nuclear Engineering and Design 328 (2018): 115-120. DOI: 10.1016/j.nucengdes.2018.01.005

[11] Bodner Research Web. Nuclearchemistry.Radioactivedecay. Availablefrom: http://chemed.chem.purdue.edu/genchem/topicreview/bp/ch23/modes.php.

[12] Colonna, N., F. Belloni, E. Berthoumieux, M. Calviani, C. Domingo-Pardo, C. Guerrero, D. Karadimos et al. "Advanced nuclear energy systems and the need of accurate nuclear data: the n_TOF project at CERN." Energy & Environmental Science 3, no. 12 (2010): 1910-1917. DOI: 10.1039/C0EE00108B

[13] Şahin, Sümer, and Yican Wu. "3.14 Fission Energy Production." (2018): 590-637.

[14] Abbasi, Akbar, and Seyedeh Fatemeh Mirekhtiary. "Risk assessment due to various terrestrial radionuclides concentrations scenarios." International journal of radiation biology 95, no. 2 (2019): 179-185. DOI: 10.1080/09553002.2019.1539881

[15] Abbasi, A. "210 Pb and 137 Cs based techniques for the estimation of sediment chronologies and sediment rates in the Anzali Lagoon, Caspian Sea." Journal of Radioanalytical and Nuclear Chemistry 322, no. 2 (2019): 319-330. DOI: 10.1007/s10967-019-06739-8

[16] Abbasi, Akbar, and Fatemeh Mirekhtiary. "137Cs and 40K concentration ratios (CRs) in annual and perennial plants in the Caspian coast." Marine pollution bulletin 146 (2019): 671-677. DOI: 10.1016/j.marpolbul.2019.06.076

[17] Abbasi, Akbar, Asley Kurnaz, Şeref Turhan, and Fatemeh Mirekhtiary. "Radiation hazards and natural radioactivity levels in surface soil samples from dwelling areas of North Cyprus." Journal of Radioanalytical and Nuclear Chemistry (2020): 1-8. DOI: 10.1007/s10967-020-07069-w

[18] Abbasi, Akbar, and Seyedeh Fatemeh Mirekhtiary. "Radiological impacts in the high-level natural radiation exposure area residents in the Ramsar, Iran." The European Physical Journal Plus 135, no. 3 (2020): 1-11. DOI: 10.1140/epjp/s13360-020-00306-x

[19] Abbasi, Akbar, and Fatemeh Mirekhtiary. "Heavy metals and natural radioactivity concentration in sediments of the Mediterranean Sea coast." Marine Pollution Bulletin 154 (2020): 111041. DOI: 10.1016/j.marpolbul.2020.111041

[20] Abbasi, A., and V. Bashiry. "Measurement of radium-226 concentration and dose calculation of drinking water samples in Guilan province of Iran." Int J Radiat Res 14, no. 4 (2016): 361-366. DOI: 10.18869/acadpub.ijrr.14.4.361

[21] Abbasi, Akbar. "A review of the analytical methodology to determine Radium-226 and Radium-228 in drinking waters." Radiochimica Acta 106, no. 10 (2018): 819-829. DOI:10.1515/ract-2018-2967

[22] Abbasi, A., and F. Mirekhtiary. "Lifetime risk assessment of Radium-226 in drinking water samples." International Journal of Radiation Research 17, no. 1 (2019): 163-169. DOI: 10.18869/acadpub.ijrr.17.1.163

[23] Abbasi, Akbar, and Fatemeh Mirekhtiary. "Some physicochemical parameters and 226Ra concentration in groundwater samples of North Guilan, Iran." Chemosphere (2020): 127113. DOI: 10.1016/j.chemosphere.2020.127113

[24] Abbasi, A., and F. Mirekhtiary. "Survey Gamma Radiation Measurementsin Commercially-used Natural Tiling Rocks in Iran." International Journal of Physical and Mathematical Sciences 5, no. 4 (2011): 561-567.

[25] Asgharizadeh, F., A. Abbasi, O. Hochaghani, and E. S. Gooya. "Natural radioactivity in granite stones used as building materials in Iran." Radiation protection dosimetry 149, no. 3 (2012): 321-326. DOI:10.1093/rpd/ncr233

[26] Abbasi, Akbar. "Environmental radiation in high exposure building materials." PhD diss., Eastern Mediterranean University (EMU)-Doğu Akdeniz Üniversitesi (DAÜ), 2013.

[27] Abbasi, A. "Calculation of gamma radiation dose rate and radon concentration due to granites used as building materials in Iran." Radiation protection dosimetry 155, no. 3 (2013): 335-342.DOI: 10.1093/rpd/nct003

[28] Abbasi, A., and M. Hassanzadeh. "Measurement and Monte Carlo simulation of γ-ray dose rate in high-exposure building materials." Nuclear Science and Techniques 28, no. 2 (2017): 20. 10.1007/s41365-016-0171-x

[29] Abbasi, Akbar. "Levels of radon and granite building materials." Radon (2017): 47.DOI:10.5772/66540

[30] Abbasi, A., and F. Mirekhtiary. "Comparison of active and passive methods for radon exhalation from a high–exposure building material." Radiation protection dosimetry 157, no. 4 (2013): 570-574. DOI: 10.1093/rpd/nct163

[31] Martin, Alan, Sam Harbison, Karen Beach, and Peter Cole. An introduction to radiation protection. CRC Press, 2018.

[32] Kaufer, B., and D. Ross. "Safety of the nuclear fuel cycle." NEA News 23, no. 1 (2005): 18-19.

Chapter 4

Does Russia Have the Possibilities to Diversify Its Export Potential to Manage the Power Engineering (For Example, Nuclear Power Development)?

Victor Kozlov

Abstract

The article analyzes the current state of power engineering, nuclear power, and their role in ensuring energy independence of Russia. According to the author, the creation of large high-tech integrated companies with active innovation state practice can bring the Russian economy to a higher level of development. To maintain Russia's leading role in the construction of nuclear power plants abroad, according to the author, it is necessary to optimize cost and terms of construction of projects, improve designs, increase scopes and quality of specialists' training, and fight corruption.

Keywords: improvement of management structure, national security, energy independence, import substitution, structural transformation, large integrated structures

1. Introduction

The situation in which turned out to be power engineering of Russia in the first decade of the twenty-first century has generated heated debate about the causes of the crisis, which turned out to be a domestic machine building, as well as ways to overcome it [1–4].

The fact that the structural transformation in the Russian machine-building complex, which took place at the time, was associated with a number of assumptions and trends largely determines the prospects for the formation of new and operation of the existing large integrated structure.

Meeting the challenges, which national engineering faces, is impossible with outraising capital in the sector, experiencing an investment "hunger." This applies to power engineering as well—a relatively prosperous industry, which was in the period of sharp decline in the domestic demand for machinery and equipment to go out of the crisis of the 1990s due to export orders with less losses than other engineering enterprises. However, the chronic underinvestment caused reducing the technical level of its production facilities.

It is necessary to focus on all resources—financial, industrial, and intellectual ones for implementation of large-scale tasks by the industry, and that in turn will require improving the management structure.

In the 1980s of the last century, the equipment supplies by power engineering provided the annual commissioning of at least 10 million kWh of electric power.

However, since 1991, there has been a sharp production decline in the industry, as evidenced by the data on manufacture of steam turbines and boilers, commissioning of generating facilities at the thermal power plants of Russia in 1990–2000, and the lack of orders for manufacturing NPP and HPP equipments [5].

"The strategy of development of power engineering of Russia," elaborated on the basis of the "Russian Energy Strategy till 2030" approved by the Government of the Russian Federation (hereinafter—the Energy Strategy), reflects the fundamental directions of development of power engineering in Russia and contains the practical measures for their effective implementation.

2. The current problems

The availability in Russia of its own effective power engineering is one of the pillars of its national security, power independence.

According to the official data, the equipment in the power industry is currently worn by almost 60%. This means that more than half of the thermal and hydropower plants operate under high risk. Given the strategic line of the state for import substitution, it is necessary to organize the process of updating the equipment in the way when orders are placed with the Russian companies, and that is possible if there are investments into domestic engineering.

Describing power engineering competitiveness, we note the peculiar feature of the domestic energy sector, which consists in the fact that almost all power plants in Russia (and CIS) are equipped with the equipment of domestic production. However, modernization and mobilization of resources in the sector can only be based on the policy of concentrating resources, pooling of capital, and formation of the effective management system. In other words, it is about solving the problem of creation of modern organizational structures.

Industrial policy in power engineering should be focused on the process of system management of its activities. Products of this sector meet the needs of other sectors of the economy as a technological component of such specific product as energy. This means that manufacture of machines and mechanisms in the power engineering industry is inseparably linked with construction and engineering works, which provide the necessary conditions for its operation.

The volume of this work is significant even in cases when the equipment is supplied for modernization of the existing facilities, rather than for equipping of new construction projects. And the technological chain of "design—manufacture—construction—installation—commissioning—operation" implies such requirements for all participants of the process of equipment commercial commissioning, the specifics of which do not allow "third-party" participants to participate in this chain (except for civil works at the facilities of power infrastructure).

Thus, the logic of the process of improving quality of the products and activities related to design, manufacture, installation, and commissioning of the equipment, as well as reducing the time for putting power units in operation requires co-operation of specialists of different branches within the same structure.

In fact, this principle was previously implemented in the framework of the branch management system. Within the USSR Ministry of Energy, about half the staff was engaged in the manufacture of power equipment, the other half—in the construction of power facilities. However, in the period of market reforms, the branch management system was destroyed, and privatized enterprises became independent market participants. But no one, even a very large factory, is able to meet the needs in power engineering products, given the specificity of these high-tech goods.

The thing is that high technologies require coordination of activities of representatives of different trades, professions, and industries. Especially in the frame of globalization when the tone is set precisely by those companies that represent a major conglomerate that combines research and production structures, as well as structures promoting products on the world market, global high-tech companies. Exactly these companies are able to bring the Russian economy to a different path of development, when export of high-tech products will be no less weighty than export of mineral resources. But none Russian factory, no matter how large it may be, is a global company; therefore, it is not competitive in world markets. Therefore, effective restructuring of production and management is necessary in the conditions of independence of market agents under insufficiency of the branch coordination system.

The process of integration of these structures, allowing to centralize development strategies and improve management and technological cooperation, will be inevitably hampered by problems of redistribution of property, as the system of securing property rights existing in Russia can be effective only if the owner is really controlling activities of all participants of the association, provided by significant proportion of ownership of assets. Given the fact that in the process of mass privatization there was no task to create an effective management system of high-tech industries, which are characterized by a high degree of co-ordination and co-operation of complex productions, formation of global companies will inevitably affect the property interests, resulting in a secondary redistribution of property. Thus, corporate conflicts are inevitable in structural transformation.

Structural transformations in the Russian machine-building complex are linked with a number of assumptions and trends, largely determining the prospects for formation of new and operation of the existing large integrated structures.

First, the experience of formation of large associated industrial structures accumulated in the Soviet times, unfortunately, was not properly developed further. However, foreign large industrial conglomerates were formed not only based on the experience of the Soviet industry but also based on the methodological basics that had been tested in the USSR.

Second, the Soviet machinery represented a hypertrophied form of simple cooperation of the universal enterprises but with huge potential unique possibilities for development. However, this potential was not even used, but, in fact, lost, in connection with the influence of accelerated privatization and breaking of the state management.

Third, the breach of economic ties and collapse of the industry management system in the process of privatization caused dominance of partnerships as a protective reaction, which are based on informal contract practices, affecting not only the process of exchange of goods but also property relations.

Fourth, the property relations established under incomplete legal frame work, regulating property relations, formed a specific model of partnership based on a system of trust relationships with contractors and the state authorities. At the same time, there were quite wide spread manifestations of economic self-interest in all aspects of the economic life, including the processes of disintegration (integration) of industrial enterprises. The new management was formed, which core competence were take over and redistribution of property.

Fifth, the dynamics of development was influenced not only by selfish holders of economic power but also by competitive strategies of foreign companies who sought to oust domestic producers from the world markets using domestic managers. However, where the owners could find a common language, a new type of engineering companies appeared, which competitiveness was high enough not only in the domestic but also in the foreign markets. As for the numerous cases of collapse of structures, those, as a rule, were associated with numerous contradictions just due to the system of partnership [6].

Thus, the corporate conflicts cannot be considered solely from a negative point of view: they are unavoidable in the process of consolidation of the technological chains belonging to different owners; moreover, replacing the owner does not always result in changing mismanagement by even more inefficient.

Implementation of major projects in the Russian power sector solves a whole range of important social and economic problems, provides employment, increases filling of regional budgets, allows to solve strategic tasks of further increase of the installed generating capacity, and increases global competitiveness of Russian equipment in particular and of Russian high-tech in general.

Large integrated structures should be active participants in the process of implementation of integrated innovations. In a sense, they act both as mechanisms of social partnership, on the one hand, expressing consolidated opinion of a large group of people involved in the manufacture of industry products, and on the other—as structures that implement decisions of the central government concerning the interests of large social groups. In addition, the large integrated structures are able to participate in development of complex innovation programs, including initiation of consideration of a number of issues by the state.

In the years 2004–2005 GAZPROM, following the recommendations of the higher state authorities acquired shares of ATOMSTROYEXPORT, actual monopoly in NPP construction abroad, and of the Incorporated Engineering Company (until 2004, a controlling stake in these companies was in the hands of K. Bendukidze, private entrepreneur). These actions of the state demonstrated that the corporation model, similar to that of the firm AREVA (France) was selected, with a majority stake owned by the state, as opposed to the corporation "General Electric," USA, which is owned by private capital [3, 7].

When Sergei Kiriyenko (1998—Prime Minister) came to ROSATOM of RF in 2005, the Agency elaborated a program of accelerated development of nuclear power in Russia [8]. This program was submitted to the government on May 18, 2006, and reported to Vladimir Putin, President of the Russian Federation, who approved the program and plan of priority measures for its implementation in June 2006.

In the development of these solutions, more than 6 billion dollars for implementation of this program were allocated in the budget. In accordance with this program, Russian NPPs should produce about 25% of the total electric power by 2030.

The nuclear industry can play an important role in solving the energy problem both in Russia and in the world. It is necessary to build and put into operation 40 GW of nuclear power units on the territory of the Russian Federation by 2030, and in other states, the Russian nuclear specialists will be able to claim the orders of 40–60 GW in the same period of time.

According to the forecasts of ROSATOM of RF by 2030, nuclear power in the world will grow up to 300–600 GW. Up to half of this promising market will be closed for external players, and the Russian nuclear specialist scan actually qualifies for 20–25% of orders (40–60 GW) of the remaining 200–300 GW of available access.

In the global nuclear fuel market, the share of Russia is now 45%. The Russian should be 50% in the markets of the USA and Canada, 42% in Europe, 35% in South Korea,

30% in Latin America, and 10% in China and Japan. To maintain its leadership, Russia has to increase capacities and implement market reforms in the sector [4].

3. Main problems that complicate the industry activities

3.1 Personnel policy and personnel training

There is no enough qualified personnel for safe operation of commissioned NPPs. The today's current system of personnel training and consolidation in the nuclear industry is clearly insufficient for its large-scale development. Working pensioners make about 25%, young workers and specialists—about 10%. Acquisition of knowledge and skills should be ahead of programs for designing and development of technologies, construction of nuclear facilities, and their commissioning. The current situation with the staff can be considered critical. With a general decline in the number of researchers (the driving force of innovation development), the share of researchers over the age of 60 years increases. The average age of leading industry experts (PhD) and university professors of "nuclear" profile is higher than the average male life expectancy in the country. Although over the past 6 years, ROSATOM has done much to address the deficiencies in training. The corporate university of 20 schools was formed on the basis of MEPI [9, 10].

3.2 Long-term construction of nuclear power plants and unreasonably overpriced construction

The long-term construction period further increases the cost and opportunities for corruption. Building "from scratch" actually takes at least 7 years, indicating shortcomings in designing and imperfect work organization.

ROSATOM stated its desire to achieve the construction time of 4–4.5 years. To do this, it is necessary to unify designs and implement innovative construction techniques based on large-block equipment supplies to construction sites.

The declared cost of 1 GW of nuclear generation has already reached $4 billion (or $4.6 billion for power unit of 1.15 GW) and continues to grow. Today the cost of NPP construction in Russia is two times higher than in China and for 30–40% higher than in Europe.

In the current environment, the economically justified cost of construction of one VVER unit of 1.15 GW capacity is not more than $2.5 billion with the construction term not more than 5 years. If ROSATOM is not able to meet these indicators, the NPP construction in the country is not competitive compared with modernization of steam turbine power units to combined cycle ones at the existing gas TPP according to the main criteria—volume of replaced gas per year during electricity generation and its net cost. The unreasonably high cost of NPP construction includes not less than 40% of the corruption component.

3.3 Electricity tariffs

The current regulated price of electricity at the Russian NPPs on the whole sale market is 3.2 US cents per kW/h (for comparison, in the USA—1.87 cents, in France and Germany—2–2.2 cents in 2008 prices). The price of electricity for economic entities in Russia is 2–3 rubles or 7–10 cents, and a new connection of consumers reaches 4.5–5 rubles or 15–17 cents (for comparison, in the USA—6.5–7.5 cents, the average price in the EU—12 cents, in China—8–9 cents). With regard to nuclear engineering, the situation is ambiguous. On the one hand, at the moment, there is possibility of producing the necessary long lead equipment for no more than three nuclear power units per year, which is obviously not enough for realization of the ambitious plans to build nuclear power plants in Russia and abroad. On the other hand, the large-scale modernization is currently carried out at the key enterprises of the energy sector of the country [6, 11].

In general, our analysis shows that in order to achieve its goals by 2030 at home and abroad, ROSATOM of RF needs to complete all the plans to modernize the machine-building enterprises in a relatively short period of time. This will allow achieving the range and scope of manufactured products to the desired level of 4–5 sets of key equipment for NPP units per year. However, given the fact that currently active negotiations are held or bidding procedures are already ongoing concerning construction of a large number of nuclear power units in a number of countries (Czech Republic, Saudi Arabia, South Africa, Kazakhstan, Nigeria, and others), more significant increase of national nuclear engineering capabilities may be required in the medium term.

To perform such wide-ranging task, it is necessary for ROSATOM of RF, at least, first of all to eliminate the disadvantages mentioned above and to pay special attention to three main areas.

The first main area is the completion of the package of administrative documents, which provide activities of enterprises of the industry and regulate the relations between the industry and the state authorities. The package of administrative documents also includes a set of more than 20 departmental purpose-oriented programs, and, of course, it includes improvement of ROSATOM structure.

The second key area is the knowledge management. It is clear that it is the high-tech industry in that all technological solutions are based on a sufficiently large block of scientific, engineering, and methodological knowledge. And various kinds of dysfunctions and failures take place without some technologization of knowledge generation, handling, and storage. For example, the knowledge is not standardized—it means that different participants of the process are based on different data. The knowledge was generated but not used in practice—hence, there is necrosis of investments in R&D. Developments were made but not commercially used—hence, there are losses in the financial sector and lack of a sufficient set of secondary developments, in which the results of major research programs are applied that have been made previously.

The third major area is the cost management. ROSATOM has been traditionally occupied in collecting data on the economy of enterprises in the industry, their processing, analysis both for planning of ROSATOM activities and in the interests of monitoring economic, financial activities of the enterprises, preparation of balance commissions, and so on. At present, this work must be carried out consistently for a radical reduction in the construction cost.

3.4 Another important

Substantiation of NPP construction calculation based on needs in power capacities, including the regional context, analysis of the grid condition in order to justify NPP connection to the general layout of power facilities by 2030.

4. Findings

The state of the national economy significantly affects the nature and methods of corporate management. This suggests the existence of specific corporate management models for each country. Thus, the formation of the corporate management national model in Russia takes place under conditions of incomplete development of the legal framework and uncertainty of ownership of privatized property, with nonexecution of the existing laws on protection of property rights and dominance of the insider control model in joint stock companies.

Thus, the problem of corporate management, which is not a purely national one, is of particular importance in the global trends. That is because the integration processes in the national high-tech sectors are characterized by the tendency of "winding-down" of internal competition in order to accumulate resources for external expansion. Therefore, the national integrated structures are involved in global competition, in which those benefit who are able to provide customers with the most comprehensive volume of services in comparison with competitors.

5. Conclusion

With regard to the real possibilities of the modern Russian nuclear power, it must be noted that over the past 15 years, five power units were constructed and put in operation abroad—in China, India, and Iran. After the long years of suspension in construction of nuclear power plants, ATOMSTROYEXPORT became the first company among its competitors, which handed over high power nuclear units complying with all safety requirements to a foreign customer. Thus, ATOMSTROYEXPORT proved to the world that the companies and organizations that make up the core of the nuclear power industry in Russia have sufficient potential and real resources to implement the most complex and demanding nuclear power projects [6].

In 2016, the assets of ASE Group companies (ATOMSTROYEXPORT)—the Engineering division of the State Atomic Energy Corporation ROSATOM, a leading player in the global market for the design and construction of nuclear energy facilities, were finally integrated. The Engineering division is well known to our foreign partners. Since its foundation, it has a reputation as an effective provider of the engineering services and has gained trust in the global market.

6. In 2018, the foreign portfolio exceeded 90 billion US dollars

By decision of the State Atomic Energy Corporation ROSATOM, the Engineering division became the Industry competence center for the management of capital construction projects.

For several years, the project management practice has been successfully implemented by ASE. The unique Multi-D technology continues to develop, being a main tool of the project management platform, which allows shortening construction time and improving labor productivity, work quality, and safety while reducing project costs. In 2016, this technology received international recognition as a winner in the WNEAWARDS competition (Le Bourget, France) presenting the "Project Management System Based on Multi-D Technologies," and that is a witness of great recognition from the world energy community. The "Multi-D® Project Management System at the Rostov NPP" won the international CEL AWARD-2016 contest in the "Megaproject" nomination, announced by FIATECH, one of the most respected industrial associations worldwide.

In addition, ASE Group companies became the first Russian company to receive an international certificate of conformity with the third competency class in the field of project, program, and portfolio management according to the International Project Management Association (IPMA Delta) model. This is another achievement internationally. Currently, certification in the field of project management according to the international IPMA standards has been passed by all top managers of the company. The division will continue to implement its strategic goals in the difficult situation of growing competition both in the NPP construction market and in the market for construction management services for the complex engineering facilities, using all resources to increase competitiveness.

Based on the successfully constructed five power units (in China, India, and Iran), the following areas of cooperation abroad are being implemented.

6.1 China

The second phase of the Tianwan NPP (TAES-2), which also includes two units with VVER-1000 reactors under the NPP-91 design, is being constructed in accordance with the General Contract for units 3 and 4 of TAES-2, signed in 2010, and entered into force in 2011. The Russian side has obligations to develop the complete engineering and operation designs of the Nuclear Island (NI) for TAES-2 units 3 and 4, providing the related services. ASE JSC also undertakes the overall technical responsibility for the design of units 3 and 4, is responsible for managing interfaces throughout the project, and provides warranty obligations.

The General Contract provides for commissioning of unit 3 in February 2018 and unit 4 in December 2018. All activities were going on schedule.

On December 30, 2017, power was launched at unit 3.

It is planned to bring the number of Russian power units in China to 8.

6.2 Iran

The implementation of the Bushehr-1 NPP project made it possible to sign the Protocol to the Intergovernmental Agreement of 08.25.1992 in 2014, which provides for the possibility to construct eight NPP units in Iran.

At the same time, on November 11, 2014, the Contract was signed under which ATOMSTROYEXPORT will construct the second and third power units of the Bushehr NPP. On September 10, 2016, the solemn laying of the "first stone" took place. The start of activities under the Contract was scheduled on December 28, 2016, when the Russian side received an advance from the Iranian customer.

During 2017, work was carried out to prepare the site. On March 14, 2001, the earthworks were started on the Bushehr-2 NPP site. On October 31, 2017, a ceremony was held to begin activities at the foundation pit of the main buildings of power unit 2. In 2018, engineering and geological surveys of the marine area and the site for spillway facilities were planned.

It was planned to coordinate the Bushehr-2 NPP design with the Customer and begin procedures related to the examination and obtaining a license for construction from the Iranian regulator. For 2018, completion of the pit for power unit 3 was scheduled, and for 2019—the "first concrete" at power unit 2. In accordance with the Contract, provisional acceptance of unit 2 is planned in 2026, of unit 3—in 2027.

6.3 India

Under the Agreement between the Government of the Russian Federation and the Government of the Republic of India on cooperation in the construction of additional nuclear power units at the Kudankulam site, as well as in the construction of nuclear power plants under the Russian designs at new sites in the Republic of India, dated December 5, 2008, the parties started the project realization plan for construction of power units 3 and 4 of the Kudankulam NPP with VVER-1000 MW reactor units each.

On October 4, 2014, the General Framework Agreement (GFA) was signed for construction of the Kudankulam NPP power units 3 and 4. In June 2017, the first concrete was poured at the second phase of the Kudankulam NPP unit 3, in October 2017—unit 4.

The planned start date for warranty operation of power units 3 and 4 is 2023 and 2024, respectively.

On June 1, 2017, ATOMSTROYEXPORT JSC and the Indian Atomic Energy Corporation signed the General Framework Agreement for the construction of the third phase of the Kudankulam NPP, and the Intergovernmental Credit Protocol necessary for implementation of the project was also signed. The Agreement provides for the construction of the third phase of the Kudankulam NPP power units 5 and 6 under the Russian design. On July 31, 2017, Contracts were signed between ATOMSTROYEXPORT JSC and the Indian Atomic Energy Corporation (IAEC) for the priority design activities and detailed design and supply of basic equipment for the third phase of the Kudankulam NPP. The planned first concrete for power units 5 and 6 is 2019 and 2020, respectively. Planned dates for the start of warranty operation of power units 5 and 6 are 2025 and 2026, respectively.

6.4 Bangladesh

On December 25, 2015, ATOMSTROYEXPORT JSC and the Bangladesh Atomic Energy Commission signed the General Contract for the construction of the Ruppur NPP consisting of 1200 MW two power units under NPP-2006 design, including a number of Appendices thereto. The signing of the General Contract was a fundamental event that allowed to start activities at the main stage of the plant construction.

In accordance with the agreement of the parties, the entry into force of the General Contract depended from fulfillment of a number of conditions. The first was signing of a credit intergovernmental agreement for the main construction period of the Ruppur NPP, then signing of Appendices to the General Contract, obtaining a license by the Bangladesh party for the NPP site and approval of the selected NPP design by the Bangladesh regulatory body.

On July 26, 2016, the Intergovernmental Agreement was signed on allocation of the state loan to finance, the main stage of the Ruppur NPP construction.

Simultaneously with the fulfillment of the conditions for the entry into force of the General Contract, in 2016 significant work was done to coordinate and prepare for signing the related integration Contracts for the Ruppur NPP project, in particular, the Contract for supply of nuclear fuel, the Contract for technical assistance for operation, service, and technical maintenance, and repair of the Ruppur NPP.

In March 2017, the parties agreed and initialed the Intergovernmental Agreement Draft on spent nuclear fuel management at the Ruppur NPP. It was planned to prepare an agreement for signing as soon as possible.

ATOMSTROYEXPORT JSC is completing the construction and installation activities at the preliminary facilities and construction and installation base. In 2016, under the General Contract, the working documentation for the main construction period was developed, as well as the materials justifying the licenses for location and construction of power units 1 and 2 of the Ruppur NPP. The first concrete is planned for 2017. Commissioning of the first unit of the Ruppur NPP is scheduled for 2022, and of the second unit—for 2023.

6.5 Hungary

History: The Hungarian-Russian cooperation in the field of nuclear energy has more than 60 years. It began in 1955 with signing of the Agreement to make a research reactor in Budapest. On December 28, 1966, the Intergovernmental Agreement was signed between Hungary and the Soviet Union on construction of the first nuclear power plant in Hungary. Currently, the Paks NPP with four VVER-440 units is successfully operating, providing more than 50% of the country's electricity.

On January 14, 2014, the Intergovernmental Agreement between Russia and Hungary was signed in Moscow on cooperation in the field of the peaceful uses of nuclear energy, which envisages the construction of two new Paks-2 NPP units.

On December 9, 2014, the Hungarian MVM Paks-2 JSC and the Russian NIAEP JSC (ASE EC JSC since December 2016) signed three Agreements regarding the construction of two NPP units with VVER-1200 Russian reactors:

- EPC—the Contract (engineering, equipment supply, and construction) for two new power units, in which the tasks for the next 10 years are fixed, taking into account the physical launch of the first unit in 2023, of the second—in 2025;

- the Contract that governs the terms of service for future power units; and

- the Contract on the conditions of long-term fuel supply.

In April 2015, the approval procedure by the EURATOM Commission for the Contract on supply of nuclear fuel for new units of the Paks-2 NPP was successfully completed.

On February 17, 2015, during the visit of the President of the Russian Federation Vladimir Putin to Hungary, the Memorandum of Understanding was signed between the State Atomic Energy Corporation ROSATOM and the Ministry of Social Resources of Hungary on training of personnel in the field of nuclear energy and related areas. According to the document, the parties will carry out cooperation in the field of education and training of personnel, educational, and scientific activities, as well as in joint educational programs in nuclear energy and related fields.

In June 2015, NIAEP JSC (ASE EC JSC since December 2016) and MVM Paks-2 JSC signed all necessary Appendices for opening financing for the EPC Contract, which stipulate the time schedule, procedure and terms of payments, and insurance conditions.

Hungary, as a member of the European Union, was obliged to carry out a total of five notification and conciliation procedures with the European Commission in connection with implementation of the Paks-2 NPP expansion project. In November 2016, the European Commission completed the expertise of the Paks-2 nuclear power plant construction project, removing all obstacles to its further development. In March 2017, construction of the new Paks-2 nuclear power units in Hungary was approved by the European Commission (EC).

The parties are developing the construction time schedule. It is planned that the license for construction of the Paks-2 NPP will be ready in 2019, and the first concrete will be poured in 2020. The peak of construction work is expected in 2021–2022. The nuclear island and the primary circuit are the responsibility of the General Contractor, while other works are carried out on procurement.

The main task of 2017 was preparation for the Paks-2 NPP construction. The scope of tasks includes preparation of engineering documentation, cooperation with suppliers, and application for a building license. As a part of the activities related to preparation of the documentation necessary to obtain licenses for the Paks-2 NPP construction, the technical design for 5 and 6 units, the preliminary safety analysis report (PSAR), and the probabilistic safety analysis reports are being prepared.

ATOMPROEKT JSC, the General Designer is completing work on the conceptual design documents that precede development of the design documentation and is also completing adaptation of the VVER-1200 base design, agreed with the Paks-2 MVM Customer, to the specific conditions of the Paks site. The process of developing chapters of the PSAR and sections of the technical design is ongoing.

The announcement of the first tender procedures has begun. Competitive information will be available on the specialized platform for ROSATOM tenders. Tender notices will also be widely published on the relevant Hungarian and international sites. All tender documentation will be posted in English. The Hungarian and other European companies can take part in procurement related to almost the entire process of the NPP construction, from design and construction to equipment supplies (except for the primary equipment that requires very specific knowledge and competencies) and related services (legal, translation, etc.).

Prospects for cooperation. In the future, when implementing the project for the Paks-2 NPP construction, all purchases of the necessary equipment and services will be carried out openly and transparently in accordance with European Union standards. As potential suppliers of equipment and services, all interested companies, including those from EU countries, can equally participate in tenders. The Russian side expects significant participation of the firms from Hungary, so that the level of localization, i.e., local industry participation, will amount to 40%.

Requirements for suppliers are different, depending on what they supply or what services they provide.

6.6 Egypt

The El Dabaa NPP (Egypt), which includes four units with VVER-1200 reactors, is being constructed in accordance with the EPC Contract, which was signed between ATOMSTROYEXPORT JSC and the Department of Nuclear Plants of the Arab Republic of Egypt on December 31, 2016 and entered into force on December 11, 2017. The project provides for construction of four power units of 1.2 GW capacity with VVER-1200 MW reactor (water-to-water power reactor) according to the Russian design. The Russian side will also assist the Egyptian partners in developing nuclear infrastructure, supply the Russian nuclear fuel for the entire life cycle of the nuclear power plant, build a special storage facility and supply containers for storing spent nuclear fuel, increase the level of localization, provide training for national personnel, and support the Egyptian partners in operation and maintenance of the El Dabaa NPP during the first 10 years of the plant's operation.

In accordance with the EPC Contract, the first power unit of the El Dabaa NPP will be commissioned in 2026.

6.7 Turkey

The Intergovernmental Agreement of the Russian Federation and Turkey on cooperation in the field of construction and operation of the nuclear power plant on the Akkuyu site in the Mersin province on the south coast of Turkey was signed on May 12, 2010.

The Akkuyu NPP project includes four power units of the Russian VVER-1200 3+ generation reactors. The capacity of each NPP unit will be 1200 MW. The design solutions of the Akkuyu NPP meet all modern requirements of the world nuclear community and established by the IAEA safety standards, the International nuclear security advisory group, and the requirements of the EUR Club.

The start of commercial operation of the Akkuyu NPP units 1–4 was tentatively scheduled for April 2023, 2024, 2025, and 2026, respectively.

6.8 Finland

On October 5, 2011, the construction site of a new nuclear power plant in Finland was announced: it will be Hanhikivi Cape in the community (municipality) of the Pyhäjoki Province, Northern Ostrobothnia (on the coast of the Gulf of Bothnia, about 100 km south of Oulu). In the media, there are various names of this plant—the Pyuhäjoki NPP, the Hanhikivi NPP, the Hanhikivi-1 NPP, but the official name is the Hanhikivi-1 NPP. It was originally planned that construction of the plant would begin in 2015, and the plant would be launched in 2020, and its maximum capacity would be 1800 MW. Initially, the negotiations were held with the companies Areva and Toshiba.

On July 3, 2013, the Finnish company Fennovoima Oy and Rusatom Overseas CJSC, a subsidiary of the Russian State Corporation ROSATOM, signed the Agreement to develop the design in order to prepare for signing of the Contract for the plant construction. It was planned that this Contract would be signed before the end of 2013. In September 2014, the Finnish government approved the NPP construction project with the participation of Russia, envisaging the use of the Russian VVER-1200 reactor. The plant should be built by 2024.

6.9 Belarus

In the Republic of Belarus, at the Ostrovets site near the city of Grodno, the construction of the Belarusian NPP consisting of VVER-1200 two power units with the total capacity of up to 2400 (2 × 1200) MW is underway. The obligations of the General Contractor are assigned to ASE JSC. It is envisaged that the Belarusian NPP is being constructed on the basis of the full "turnkey" responsibility of the General Contractor. The "NPP-2006" design, the General Designer is ATOMPROEKT JSC, was chosen for construction of the first Belarusian NPP.

The Belarusian NPP design complies with all international standards and IAEA recommendations and is characterized by the enhanced safety characteristics, technical, and economic indicators.

The main advantages of the Russian design are a high degree of security provided through the use of the independent channels of active and passive safety systems, the melt trap, and other systems. Unit 1 of the plant is planned to be commissioned in 2019, unit 2—in 2020.

The great prospects imply an even greater responsibility. The previous story, now of the ASE Group companies, allows to hope for successful implementation, I am not afraid of the word, of the grandiose tasks, by the nuclear power industry of Russia!

Author details

Victor Kozlov
Russian University of Economics Named After Plekhanov, Moscow

*Address all correspondence to: kozlvik@mail.ru

References

[1] Magazine. Economist. 2004;7:24

[2] Kozlov VV. Corporate Structures. 2004

[3] Paleotyp M. Magazine. Problems of Forecasting. 2005;3:45

[4] Kozlov V. Personal training in the field of energy and its role in the economical safety of Russia

[5] Nigmatulin B. First Deputy Director of the Institute of Natural Monopolies. Open Letter to Kiriyenko S.V., CEO of SC ROSATOM. Available from: http://www.atom44.ru/component/content/article/7-novosti-atomnojj-ehnergetiki/45-nigmatulin

[6] Atomic Expert. The Global Nuclear Fuel Market—Supply and Demand 2013-2030. 2013. Available from: http://atomicexpert.com/sitemap [Accessed: 24.02.2014]

[7] Magazine. Innovation and Investments, Moscow, #2. 2014

[8] Rosatom of RF annual report for 2013

[9] Kozlov V. Russian nuclear energy abroad. Monograph. 2018

[10] Available from: http://www.Atomstroyexport.ru

[11] Rosatom engineering division—group of the companies ASE, annual report 2018

Chapter 5

Fast-Spectrum Fluoride Molten Salt Reactor (FFMSR) with Ultimately Reduced Radiotoxicity of Nuclear Wastes

Yasuo Hirose

Abstract

A mixture of NaF-KF-UF_4 eutectic and NaF-KF-$TRUF_3$ eutectic containing heavy elements as much as 2.8 g/cc makes a fast-spectrum molten salt reactor based upon the U-Pu cycle available without a blanket. It does not object breeding but a stable operation without fissile makeup under practical contingencies. It is highly integrated with online dry chemical processes based on "selective oxide precipitation" to create a U-Pu cycle to provide as low as 0.01% leakage of TRU and nominated as the FFMSR. This certifies that the radiotoxicity of HLW for 1500 effective full power days (EFPD) operation can be equivalent to 405 tons of depleted uranium after 500 years cooling without Partition and Transmutation (P&T). A certain amount of U-TRU mixture recovered from LWR spent fuel is loaded after the initial criticality until U-Pu equilibrium but the fixed amount of ^{238}U only thereafter. The TRU inventory in an FFMSR stays at an equilibrium perpetually. Accumulation of spent fuel of an LWR for 55 years should afford to start up the identical thermal capacity of FFMSR and to keep operation hypothetically until running out of ^{238}U. Full deployment of the FFMSR should make the entire fuel cycle infrastructures needless except the HLW disposal site.

Keywords: fast-spectrum fluoride molten salt reactor, high-level radioactive waste, structure of fuel salt, density of fuel salt, redox potential control, freezing behavior of fuel salt, selective oxide precipitation process, front-end processing dedicated to the MOX spent fuel, nuclear fuel cycle and associated wastes

1. Introduction

Almost but a few would recognize the relation between fossil fuel burning and the global greenhouse issue. However many of them tend to be in favor of expensive and inefficient but immediately harmless renewable energy than existing nuclear. A major barrier to persuade con-nuclear elements is the nuclear waste issue which should have directly associated with Pu production for the traditional strategy to close fuel cycle to ensure national energy security by using the liquid metal fast breeder reactor (LMFBR). In coping with this circumstance, resolutions to address both issues, i.e., decreasing radiotoxicity of the high-level radioactive waste

(HLW) and sustaining energy using the molten salt reactor (MSR) technology, have been expected.

The MSR technology was developed and culminated by successful operation of the molten salt reactor experiment (MSRE) and conceptual design of the molten salt breeder reactor (MSBR) in Oak Ridge National Laboratory (ORNL) by 1975 to make thorium as a naturally available fuel usable in addition to uranium.

Many attempts have been made to realize breeder reactors based on the U-Pu cycle using molten salt fuels; however such endeavors had been limited in the chloride salts, because of the feasibility to obtain high enough energy of neutron flux [1].

In addition to the predicted solubility data from thermodynamic calculations [2], recently actualized high solubility data of various fluorides of actinides and lanthanides specifically in the LiF-NaF-KF eutectic mixture (traditionally named as FLiNaK) [3–6] reportedly could allow utilizing the high enough energy neutrons for the U-Pu breeding cycle aided by a high heavy element inventory (2.8 t/m^3) and a small neutron moderating capability [7].

It was reported that a system nominated as a 3.2 GWt U-Pu fast-spectrum molten salt reactor (U-Pu FMSR) of 21.2 m^3 core volume (31.8 m^3 total primary system volume) starting from 68.5 tons of uranium with 15 tons of plutonium solving in FLiNaK to reach an equilibrium state after 10 years with an online chemical processing in which the conversion ratio (CR) became positive with inventory of 68.6 tons uranium, 20.9 tons plutonium, and 1.4 tons miner actinides did not need fissile material in feeding and consumed ^{238}U only [7].

It does not intend to reduce doubling time but can breed fissile to guarantee stable operation without a blanket. It does not deliberately decrease TRU but confines them into the reactor core and isolates them from improper uses indefinitely. It does continuously renew fissionable actinides by the metabolic function with an online processing and produce nearly actinide-free fission product streams to be wasted. This implies that fluoride molten salt reactor technology is being available based upon U-Pu breeding cycle to afford reasonable approach to global task addressed to sustaining natural resources, decreasing stockpile of plutonium as well as depleted uranium, relieving radioactive waste burden from the use of nuclear energy, achieving complete nonproliferation, inheriting safety characteristics of the liquid fuel, and establishing complete stand-alone system associating with only the waste disposal facility.

The author and associates have successfully performed a follow-up calculation not only for FLiNaK but also NaF-KF-UF_4 system as a matrix of the fuel salt [8, 9]. Their efforts have borne a fruit as a nuclear reactor plant using a mixture of NaF-KF-UF_4 fertile and NaF-KF-$TRUF_3$ fissile as the fuel salt incorporated with designated online chemical processes based upon the oxide selective precipitation process with extremely low heavy element released to the environment which was nominated as the fast-spectrum fluoride molten salt reactor (FFMSR) [10–12].

2. Preliminary survey and study

2.1 Is FLiNaK the best choice as the matrix for a liquid fuel?

Composing the fuel salt for a thermal reactor such as the Molten Salt Breeder Reactor (MSBR) had nothing to do with solubility. The fertile salt 0.72LiF-0.16BeF_2-0.12ThF_4 had a unique phase relationship in which liquidus was constant at 500°C during ThF_4 content was varied between 10 and 20 mol%. The fissile salt $^{233}UF_4$ was not dissolved in the fertile salt, but displaced $^{232}ThF_4$, as they had the same monoclinic crystal structure. During the freezing process, almost 75% of the fuel salt was solidified at 500°C as the same composition as the liquid phase. Eventually $0.47LiF\text{-}0.515BeF_2\text{-}0.015ThF_4$ containing very small amount of $^{233}UF_4$ solidified as eutectic at 370°C [13]. This freezing process was evaluated as nuclear criticality safety in the fuel salt drain tank, and it became as a basis of the technological feasibility.

In serious attempt to use hexagonal PuF_3 as a fissile instead of monoclinic $^{233}UF_4$ in $0.72LiF\text{-}0.16BeF_2\text{-}0.12ThF_4$, it had been treated as the solubility of PuF_3. The term "solubility" has been used as a convenient synopsis of "liquefied fraction" on the phase diagram [14]. However they are not the same exactly because "solubility" is defined as the mole fraction of solute in solvent, while "liquefied fraction" is defined as the fraction in total mole value.

The elaborated solubility measurements in FLiNaK by Russian scientists [3–6] should have been more appropriately respected if they had made the chemical composition of alkali fluoride matrix of liquefied samples analytically quantified instead of their customary practice in which the matrix had been always assumed as FLiNaK, even if they have found no UF_4 or PuF_3 but $2KF\text{-}UF_4$, $7KF\text{-}6UF_4$, KPu_2F_7, $KPuF_4$, and $NaPuF_4$ in the solidified residue by the X-ray diffractometric analysis.

The author tries to interpret the solubility of UF_4 and PuF_3 in the FLiNaK by producing liquefied components at respective temperatures as shown in **Table 1** based upon the material balance referring from relevant phase diagrams in **Figure 1** [15] and **Figure 2** [16]. The red line in each ternary diagram which starts from the actinide fluoride corner, passes through the eutectic point, and ends in the alkali fluoride edge represents the actinide concentration in a fixed matrix composition.

The increasing process of liquefied fraction consists of two types, firstly composing compounds at the eutectic temperature and secondly increasing content of liquefied fraction according to rising temperature. Alkali fluoride compounds of UF_4 have a wider range of liquid zone than those of PuF_3 in the relevant phase diagrams.

	Matrix	Melt Temp. °C	Compounds in Liquid Phase	Eutectic Temp. °C	Melt Temp. °C	UF_4 Mol%	PuF_3 Mol%
FLiNaK-UF4			UF_4		1035		
			$0.259LiF\text{-}0.259UF_4$		735	50.54	
			$0.206NaF\text{-}0.115KF\text{-}0.260UF_4$		735	46.21	
	0.420KF	865	$0.420KF\text{-}0.503UF_4$	735		42.89	
	0.259LiF-0.420KF	492	$0.259LiF\text{-}0.096UF_4$	490		19.88	
	0.465LiF-0.115NaF-0.420KF	454	$0.206LiF\text{-}0.115NaF\text{-}0.152UF_4$	445		13.19	
FLiNaK-PuF3			PuF_3		1396		
			$0.161LiF\text{-}0.044NaF\text{-}0.210PuF_3$		726		36.35
			$0.304LiF\text{-}0.420KF\text{-}0.340PuF_3$		726		29.32
	0.07NaF	900	$0.071NaF\text{-}0.021PuF_3$	726			19.80
	0.161LiF-0.115NaF	760	$0.161LiF\text{-}0.044NaF\text{-}0.058PuF_3$	685			18.43
	0.465LiF-0.115NaF-0.420KF	454	$0.304LiF\text{-}0.420KF\text{-}0.168PuF_3$	513			14.38
FLiNaK-UF4-PuF3			PuF_3		1396		
			UF_4		1035		
			$0.259LiF\text{-}0.259UF_4$		735	42.00	
			$0.420KF\text{-}0.420PuF_3$		726		36.10
			$0.420KF\text{-}0.320PuF_3$		700		26.11
			$0.206LiF\text{-}0.115NaF\text{-}0.248UF_4$		666	28.64	
	0.420KF	856	$0.420KF\text{-}0.225PuF_3$	619			18.36
	0.259LiF-0.420KF	492	$0.259LiF\text{-}0.096UF_4$	490		19.88	
	0.465LiF-0.115NaF-0.420KF	454	$0.206LiF\text{-}0.115NaF\text{-}0.152UF_4$	445		13.19	

Table 1.
Interpretation of solubility upon accumulated liquefied compounds.

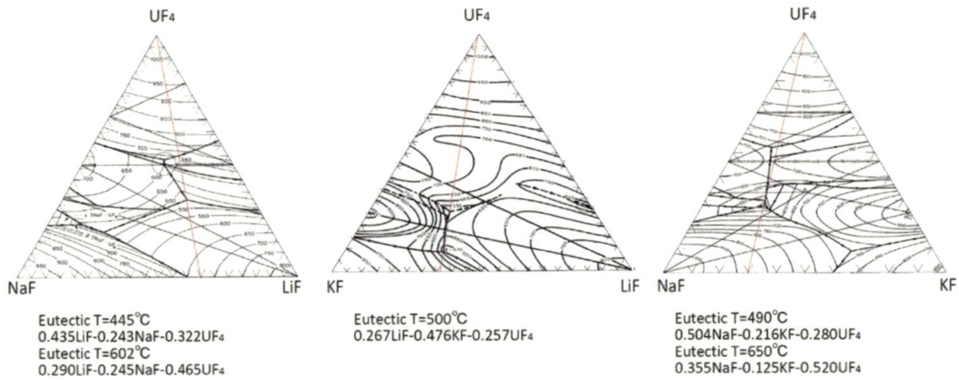

Figure 1.
Phase diagrams for LiF, NaF, KF, and UF_4 system [15].

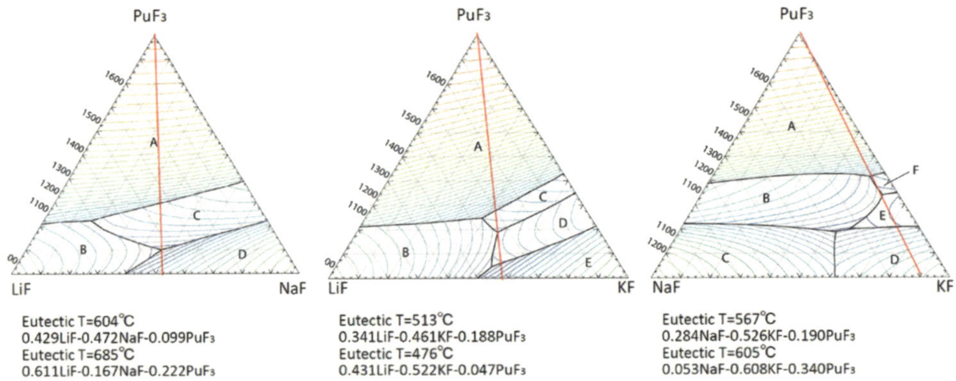

Figure 2.
Phase diagrams for LiF, NaF, KF, and PuF_3 system [16].

Coexistence of UF_4 and PuF_3 obviously competes each other in the first mechanism. The eutectic formation at lower temperature should have the priority.

These results would be summarized as:

1. The liquefied mixture of FLiNaK and heavy metal fluorides is not a solution.

2. KF (mp = 865°C) might have been temporally solidified prior to producing 0.445KF-0.555UF_4 (735°C) during the ascending temperature process in the solubility measurement of UF_4.

3. KF (mp = 865°C) and NaF (mp = 900°C) might have been temporally solidified prior to producing 0.651KF-0.349PuF_3 (619°C) or 0.772NaF-0.228PuF_3 (726°C) during the ascending temperature process in the solubility measurement of PuF_3.

4. The saturated FLiNaK solution of UF_4 and PuF_3 is elucidated as the mixture of three types of alkali fluoride compound assumed as 0.321 (0.435LiF-0.243NaF-0.322UF_4)-0.241 (0.730LiF-0.270UF_4)-0.438 (0.651KF-0.349PuF_3) with liquidus temperature of 619°C and solidus temperature of 445°C.

5. The liquidus temperature of the FLiNaK mixture might be substantially higher than that of solvent. Any physical favorable properties of FLiNaK should have not been directly attributed to the fuel salt.

2.2 Alternative choice to prepare the liquid fuel

Taking the lessons learned, the liquid fuel has to be a mixture of fertile salt and fissile salt both frozen into eutectic phases. Extensive numbers of phase diagram, which show the relationship between the variation of compositions and the liquidus temperature of mixtures, for alkali fluoride systems containing UF_4 and for those containing PuF_3 have been defined. The eutectic temperature means that nothing but liquid is stable over this temperature and that nothing but solid is stable under this temperature. The eutectic compositions and temperatures for the alkali fluoride systems containing UF_4 and PuF_3 are listed in **Table 2**.

There are various candidates for the combination of fertile salt and fissile salt as shown in **Table 3**. Technologically the liquidus temperature is preferably as low as possible. The lower heavy metal content of a component could imply higher liquidus temperature apart from the indicated eutectic temperature.

The author is particularly interested in the fuel system consisting of NaF-KF-UF_4 and NaF-KF-PuF_3 which do not contain enriched ^{7}LiF for economic as well as technological reasons associated with tritium control and irradiation defects after being solidified. If there might be a particular reason to contain LiF in the fuel, it is decreasing viscosity.

It is revealed that this combination can provide 0.35NaF-0.29KF-0.28UF_4-0.08PuF_3 composed of mixing 0.762 (0.504NaF-0.216KF-0.280UF_4) and 0.238 (0.053NaF-0.608KF-0.340PuF_3) at the liquidus of 605°C and the solidus of 490°C. This means that nothing but liquid is stable at 605°C or higher and nothing but solid is stable at 490°C or lower according to the phase diagrams **Figures 1** and **2**.

Alkali fluoride with UF_4 [15]			Alkali fluoride with PuF_3 [16]		
Compositions	**Molecular ratio**	**ET***	**Compositions**	**Molecular ratio**	**ET***
LiF-UF_4	0.730–0.270	490	LiF-PuF_3	0.798–0.212	745
			LiF-UF_4-PuF_3	0.733–0.257–0.010	484
NaF-UF_4	0.785–0.215	735	NaF-PuF_3	0.779–0.221	726
	0.720–0.280	623			
	0.440–0.560	680			
KF-UF_4	0.850–0.150	618	KF-PuF_3	0.651–0.349	619
	0.615–0.385	740			
	0.460–0.540	735			
LiF-NaF-UF_4	0.600–0.210–0.190	480	LiF-NaF-PuF_3	0.429–0.472–0.099	604
	0.350–0.370–0.280	480		0.611–0.167–0.222	685
	0.435–0.243–0.322	445			
	0.245–0.290–0.465	602			
LiF-KF-UF_4	0.331–0.589–0.080	470	LiF-KF-PuF_3	0.431–0.522–0.047	476
	0.267–0.476–0.257	500		0.341–0.471–0.188	513
NaF-KF-UF_4	0.293–0.622–0.085	650	NaF-KF-PuF_3	0.285–0.528–0.187	567
	0.504–0.216–0.280	490		0.053–0.607–0.340	605
	0.355–0.120-0.520	650			

**Eutectic temperature.*

Table 2.
Alkali fluoride eutectic mixture containing UF_4 or PuF_3.

Case	Fertile salt (eutectic temp.)	Fissile salt (eutectic temp.)
Li	0.730LiF-0.270UF$_4$ (490°C)	0.788LiF-0.212PuF$_3$ (745°C)
Na	0.720NaF-0.280UF$_4$ (623°C)	0.779NaF-0.221PuF$_3$ (726°C)
K	0.850KF-0.150UF$_4$ (618°C)	0.651KF-0.349PuF$_3$ (619°C)
Li-Na	0.435LiF-0.243NaF-0.322UF$_4$ (445°C)	0.611LiF-0.167NaF-0.222PuF$_3$ (685°C)
Li-Na-K	0.435LiF-0.243NaF-0.322UF$_4$ (445°C)	0.341LiF-0.461KF-0.188PuF$_3$ (513°C)
Li-K	0.267LiF-0.476KF-0.257UF$_4$ (500°C)	0.341LiF-0.461KF-0.188PuF$_3$ (513°C)
Li-K-Na	0.267LiF-0.476KF-0.257UF$_4$ (500°C)	0.611LiF-0.167NaF-0.222PuF$_3$ (685°C)
Na-K	0.504NaF-0.216KF-0.280UF$_4$ (490°C)	0.053NaF-0.608KF-0.340PuF$_3$ (605°C)

Table 3.
Candidates for the combination of fertile salt and fissile salt.

2.3 Density of alkali fluoride mixture with heavy metal fluoride

The density of a liquid mixture has been customarily obtained as a reciprocal of a weighted average of molecular volume of components; though this procedure worked satisfactorily during the MSRE and MSBR project in ORNL [17], concurrently it has been recognized that the results might be significantly erroneous without pertinent information about the respective components, e.g., liquid UF_4 or PuF_3. If the components would compose a complex compound, e.g., $2KF + UF_4 \rightarrow K_2UF_6$ or $3KF + PuF_3 \rightarrow K_3PuF_6$, it might cause a serious deviation from linearity.

Since most molten salt reactors considered during the early stages of MSR project in ORNL were thermal or epithermal, the fluorides of lithium, beryllium, sodium, and zirconium have been given the most serious attention for the carrier salt of liquid fuels. However some alkali fluoride mixtures including potassium with UF_4 were also investigated in ORNL during the earlier stage of MSR project although details had been classified [18]; however the density data were perceived as not from additivity calculation as listed in **Table 4**.

However it seems that the density of listed mixtures is approximately expressed by a couple of second-order approximate least square functions according to UF_4 molar concentration, one for binary systems and another for ternary (or pseudo-ternary) systems, regardless of alkali fluoride matrix as shown in **Figure 3**.

Based upon the density data for solid UF_4, UF_3, PuF_4, and PuF_3, i.e., 6.72, 8.97, 7.0, and 9.32 g/cm^3 at the room temperature [19], it is hypothetically assumed that PuF_3 can be substituted by 1.389 molecules of UF_4 and UF_3 by 1.335 molecules of UF_4 in the sense of density effect. The average temperature coefficients were reported as 0.0008/°C in the range of 0–4 mol% and as 0.0011/°C in the range higher than 22 mol% [18].

This procedure to estimate the density of fuel salts with substantially high concentration of actinides became a major breakthrough in the whole study; however it should be experimentally verified further (**Table 5** [19]).

2.4 Implication of density of the liquid fuel in the feasibility of reactor

2.4.1 Effect of density on conversion of inventories to concentrations

The physical feasibility of the U-Pu FMSR was independently verified by us in the sense of heavy element inventory with small deviations [8]; however there have been drastic differences in mol% concentrations of UF_4 and PuF_3 to provide the

Composition of salt				MP	Liquid density	Liquid viscosity	Specific heat at 700°C	Thermal conductivity
Li	Na	K	U	°C	(g/cc)(T:°C)	(Cp)	(cal/g-deg)	(W/m-K)
	60	40		710	2.40-0.00060 T	–	–	–
60	40			652	2.42-0.00055 T	4.66 (600°C)	0.58	–
50		50		492	2.46-0.00068 T	4.75 (600°C)	0.44	–
46.5	11.5	42		454	2.53-0.00073 T	4.75 (600°C)	0.45	4.53
72.5			27.5	490	6.11-0.00127 T	12.1 (700°C)	–	–
	66.7		33.3	623	5.51-0.00130 T	16.3 (600°C)*	0.21	–
	50		50	680	6.16-0.00107 T	–	–	–
		45	55	735	6.07-0.00115 T	–	–	–
38.4	57.6		4	645	2.95-0.00770 T	3.5 (700°C)	0.53*	-
33	45		22	506	4.50-0.00101 T	–	0.26	–
48		48	4	560	2.75-0.00073 T	3.2 (700°C)	0.38	–
	48.2	26.8	25	558	4.54-0.00110 T	9.8 (700°C)	0.23	–
	46.5	26	27.5	530	4.70-0.00115 T	17.3 (600°C)	0.23*	0.87
	50	20	30	575	4.78-0.00104 T	10.0 (700°C)	0.22	–
	35	20	45	708	5.60-0.00116 T	–	–	–
44.5	10.9	43.5	1.1	452	2.65-0.00090T*	4.61 (600°C)*	0.44*	4
45.3	11.2	41	2.5	490	2.67-0.00072 T	5.10 (600°C)*	0.38	–
44.7	11	30.3	4	560	2.80-0.00074 T	5.35 (600°C)	0.41	–

**Explicitly marked as experimental value.*

Table 4.
Some physical properties of alkali fluorides containing UF_4 [18].

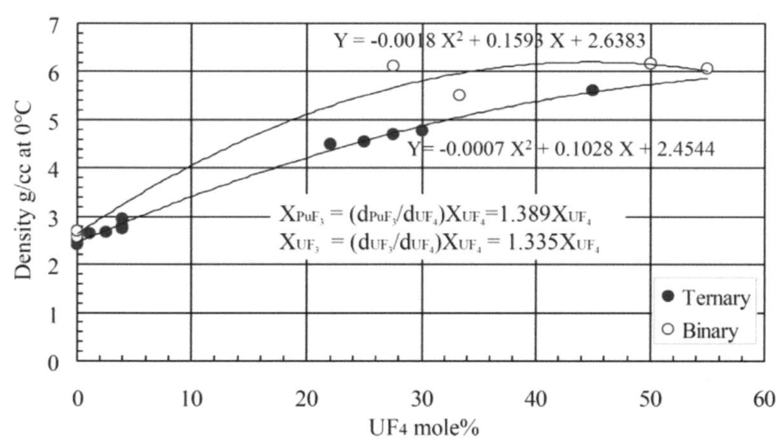

Figure 3.
Density of alkali fluorides containing UF_4 [18].

required inventory as shown in **Table 6**. We have learned that the reported work [7] have applied density of the fuel salt as 5.32 g/cc at 680°C derived from the weighted average process of molecular volume for 0.704 (FLiNaK)-0.21UF_4-0.067PuF_3-0.0045MaF_3-0.014FP [20], while it was 3.86 g/cc according to our procedure.

	ThF_4	UF_4	PuF_4	UF_3	PuF_3	CeF_3
Free energy of formation at 1000 K (kcal/F atom)	−101	−95.3	−86.0	−99.9	−104.3	−118
Melting point (°C)	1111	1035	1037	1495	1425	1637
Crystal structure*	M	M	M	H	H	H
Density (g/cc) at 20°C	5.71	6.72	7.0	8.97	9.32	6.16

**M, monoclinic; H, hexagonal.*

Table 5.
Comparison of properties of PuF_3 with ThF_4, UF_4, PuF_4, UF_3, and CeF_3 [19].

	U-Pu FMSR [7]	Our work [8]		
Fuel salt	**UF_4-PuF_3 in FLiNaK**	**UF_4-PuF_3 in FLiNaK**	**PuF_3 in NaK-KF-UF_4**	**$TRUF_3$ in NaF-KF-UF_4**
Power, MWth	3200	3200	3200	3200
Reactor core H/R ratio, h/r	1.85	1.85	1.85	1.85
Reactor core volume, m^3	21.2	21.2	21.2	21.2
Specific power, W/cm^3	150	150	150	150
Average neutron flux, $cm^{-2}\ s^{-1}$	$\sim 10^{15}$			
Initial fuel loading, U/Pu/MA, ton	68.5/15/−	72.1/16.1/−	71.3/15.6/−	71.3/17.1/2.1
Equil. fuel loading, U/Pu/MA, ton	68.6/20.9/1.4	71.9/20.3/1.2	71.2/19.2/1.3	71.2/19.6/1.2
Fuel salt density, g/cc, at 680°C*	5.32			
Fuel salt density, g/cc, at 680°C**	3.862	4.442	4.343	4.358
k_{eff} in equil. state	1.008	1.008	1.008	1.008
k_∞ in equil. state	1.044	1.054	1.051	1.052
Temperature coefficient	$-2.4\text{–}10^{-5}$	$-8.0\text{–}10^{-5}$	$-7.6\text{–}10^{-5}$	$-7.3\text{–}10^{-5}$

**Weighted average of molecular volume.*
***Interpolated from the ORNL data.*

Table 6.
Results of the follow-up calculations.

The calculated molar concentration of the heavy elements in the fuel salt is inversely proportional to the density of the fuel salt for the identical inventories. The nuclear characteristics rely on the heavy metal inventory; however the phase relationship and chemical/hydrothermal characteristic solely rely on molecular concentration of heavy metal fluorides. Accordingly the fuel composition for which we have to examine the technological feasibilities should be 0.612 (FLiNaK)-0.290UF_4-0.098$TRUF_3$ instead of 0.704 (FLiNaK)-0.21UF_4-0.067PuF_3-0.0045MaF_3-0.014FP.

Establishing the standard process to evaluate reliable density value of the fuel salt is an indispensable step of research and development work of the molten salt reactor technology particularly when it is across multiple research parties.

2.4.2 Deviation of density due to UF_3 formation

The physical calculations up to now for the FFMSR are performed for the fuel salt having chemical composition as.

$$0.350\text{NaF} - 0.290\text{KF} - 0.280\text{UF}_4 - 0.080\text{PuF}_3 \; (d_{928\text{K}} : 4.684\, g/\text{cc}) \quad (1)$$

However if $[UF_3]/[UF_4]$ ratio should have been kept at 5% for the redox buffer control as will be discussed in Section 2.5.2, the chemical composition might have been altered as.

$$0.350\text{NaF} - 0.290\text{KF} - 0.267\text{UF}_4 - 0.013\text{UF}_3 - 0.080\text{PuF}_3 \; (d_{928\text{K}} : 4.714\, g/\text{cc}) \quad (2)$$

This unique temperature unrelated factor ($\pm 0.06\%$ of fuel density) on the reactivity should be evaluated accordingly.

2.5 Challenges for realization of FFMSR

2.5.1 Characteristic arrangement for the unmoderated MSR

The authors have never dared to realize molten salt fast reactors for burning TRU, unless we could have seen a tank-within-tank layout proposed by Forsberg [21] and reproduced in **Figure 4**, to ensure characteristic safety of the unmoderated MSR based on the technology for the fluoride high-temperature reactor (FHR).

A unique criticality safety challenge associated with unmoderated MSR is that criticality can occur if the fissile materials leak from the system and come near the neutron moderators, such as concrete. This has to exclude the "catch pan" arrangement to transfer gravitationally the spilled fuel material into the drain tank, which has been traditionally adapted by graphite-moderated MSR.

The combination of the direct reactor cooling system (DRACS), the pool reactor auxiliary cooling system (PRACS), and the buffer-salt pool which includes drain tanks in the bottom and is located in the underground silo can accommodate the decay heat removal and criticality issues under the design basis as well as the beyond-design-basis accident, even including the outer vessel failure.

2.5.2 Redox control of FFMSR

UF_4 molecule in a liquid fluoride mixture intrinsically oxidizes to dissolve Cr as the most vulnerable constitution of the specifically developed structural material Hastelloy N to result in CrF_2 and to form UF_3 molecule. This challenge to be

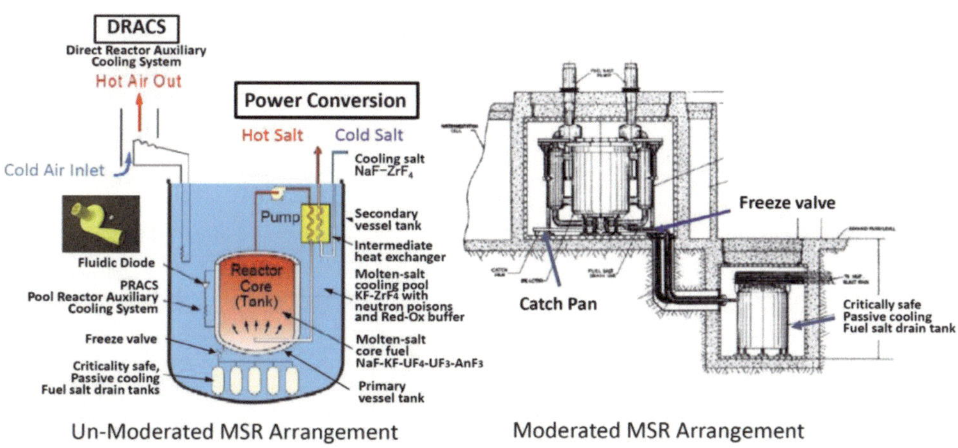

Figure 4.
Comparison between unmoderated and moderated arrangement [21].

addressed for a UF_4-fueled molten salt reactor was overcome by keeping U(IV)/U(III) ratio no less than 100 with constant monitoring of CrF_2 concentration [22].

From 1965 to 1969, a successful operation of the MSRE proved that the fission of $^{235}UF_4$ as well as of $^{233}UF_4$ made the fuel salt moderately oxidizing as previously suggested and proven that the absence of metallic uranium deposition or uranium carbide formation incidence due to successive fissioning. The U(IV)/U(III) ratio could be maintained within the projected range by periodic dissolution of beryllium metal bar suspended in the pump bowl. During the post-MSRE work, it was found that the significant intergranular cracks due to the presence of fission product tellurium could be suppressed by adjusting the U(IV)/U(III) ratio no higher than 70 [22].

In 1968, Thoma [19] described that no significant differences were believed to exist in the yield or chemistry of the principal species of fission products which would result from the incorporation of PuF_3 in MSR fuels and then the use of a tri-fluoride solute should result in a cation excess and should cause the fuel solution to generate a mild reducing potential, because he had confirmed that the fission of $^{235}UF_4$ fuel consuming ~0.8 is equivalent to UF_3 per gram atom of fissioned uranium.

In 1994 Toth [23] ratified Thoma's perception [19] made in 1968 regarding the effect of PuF_3 fission on redox potential of the fuel salt however with strong warning that further investigations should be required if Pu fuels were used in future designs.

In November 2017, the ORNL has made an official presentation to the US-NRC staff [24] that the fission of PuF_3 releases three fluorine ions, while the fission products require more than three, and thus there will be a fluorine ion deficit with net reducing conditions without showing fission product yield data or chemical status of fission products. The ORNL traditionally has ignored the fact in which fission of ^{239}Pu yields much more rare metals and much less zirconium than those of ^{235}U or ^{233}U which could decrease the required fluorine ions substantially.

The author solicited Dr. Shimazu [25] to take a positive approach to certify the new redox potential control paradigm using the newest computation practice and elucidated free fluorine yield data for $^{233}UF_4$, $^{235}UF_4$, and $^{239}PuF_3$ per unit fission as well as per unit power output under both thermal neutron (MSRE) and fast neutron (FFMSR) environment assuming that the chemical behavior of fission product in molten fluoride environment is identical as evaluated for ^{235}U fission by Baes [26] as shown in **Table 7**.

It was informed by the study [27] with molten LiF-BeF_2-ThF_4 (75-5-20 mol%) salt mixture fueled by 2 mol% of UF_4 and containing additives of Cr_3Te_4, including 250-h tests with exposure of nickel-based alloy specimens at temperatures from 700 to 750°C and under mechanical loading, that there were no traces of tellurium intergranular cracking on specimens in the fuel salt with [U(IV)]/[U(III)] ratio from 20 to 70 and no nickel-uranium intermetallic film on the specimens with fuel salts characterized by the ratio larger than 3, as shown by the acceptable redox voltage range in **Figure 5** [27, 28].

Fissionable Materials	Mole-F/Fission		Mole-F/MWt-y	
	Fast	Thermal	Fast	Thermal
$^{233}UF_4$	0.65	0.80	1.13	1.14
$^{235}UF_4$	0.80	0.80	1.35	1.36
$^{239}PuF_3$	0.60	0.60	1.02	1.09

Table 7.
Free fluorine production rate per fissioning in liquid fluoride fuel [25].

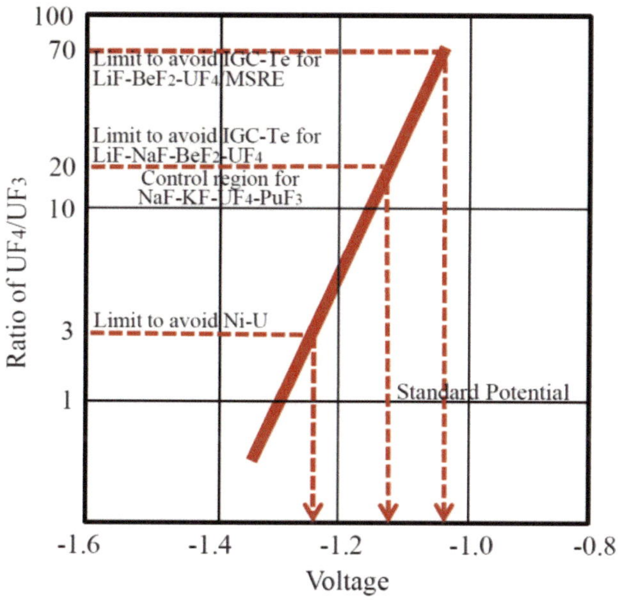

Figure 5.
Dependence of the redox potential on UF_4/UF_3 ratio [27, 28].

$NaF\text{-}KF\text{-}UF_4$ compound might be less Lewis basic than $LiF\text{-}BeF_2\text{-}ThF_4$; however it will require to make the [U(IV)]/[U(III)] ratio at least 20. This means that UF_3 has to be kept as the redox buffer at 1.33 mol% while total uranium fluoride at 28.1 mol%.

2.6 Operational behavior of FFMSR

2.6.1 Effect of U inventory on reactor physical properties

Neutronic calculations were made taking originally proposed configurations of the reactor (core height/radius ratio, 1.85; core volume, 21.2 m^3; primary circuit volume, 31.8 m^3) and the power output (3.2GW_{th}) the same as Ref. [7], but other factors, such as the actinide isotopic composition (45,000MWD/t-U in BWR, 5 years cooling), neutron leakage (with 30 cm steel reflector), the fuel temperature (627°C), the salt cleanup and makeup condition, etc., were discretely specified to give verified number of heavy element masses and concentrations in the fuel salt to give designated reactivity (k_{eff} = 1.007) from the start up to the equilibrium state (40 years).

Operational features are characterized by annual feed/breed balance of fissile material as TRU over four zones under a constant U inventory which can be maintained by an appropriate makeup. The effect of U inventory in three levels on TRU feed/breed balance is evaluated in which fuel salt cleaning started after 300 effective full power days (EFPD) with an interval of 300 EFPD and illustrated in **Figure 6**. The larger inventory of U requires larger amount of initial fissile inventory but smaller amount of supplement; however the peak annual supplement is less dependent on the initial U charge. U inventory of 61.4 tons is the lowest threshold limit to make breeding break-even possible, while that of 71.5 tons can provide as much as 100 kg TRU of annual breeding; however it is the highest threshold limit by U content acceptable by a relevant fuel salt.

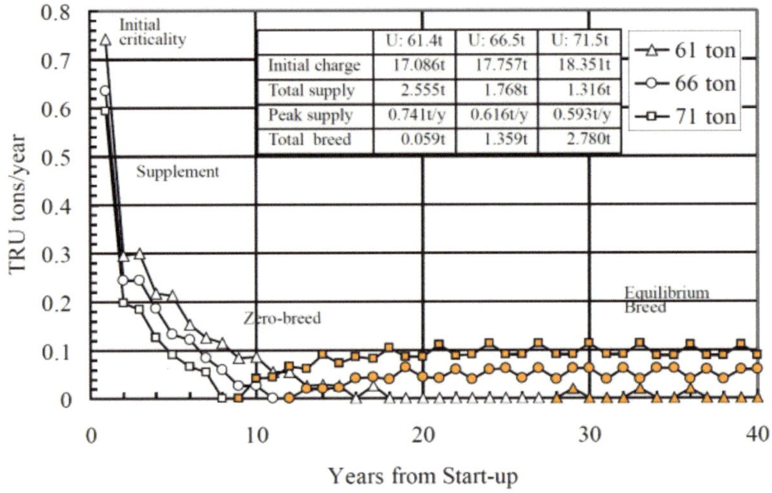

Figure 6.
Effect of initial U charge on the feed/breed balance.

2.6.2 Effect of fuel salt cleaning interval on reactor physical properties

Three hundred EFPD and 1500 EFPD of the fuel salt cleaning interval are evaluated both for an identical initial charge of the fuel salt composition (U: 71 t) as shown in **Figure 7**. No chemical cleaning but only makeup of TRU was made during the designated initial interval. The longer interval requires larger amount of fissile material supplement; however the peak annual supplement is less dependent on the extension of cleaning interval. A longer interval makes the cleaning volume smaller but nevertheless total makeup larger; however the cost of facility is specifically determined by the peak annual makeup value.

The operation of an FFMSR with 1500 EFPD of fuel salt cleaning interval is assumed as barely providing a steady and sustaining operation with an appreciable breeding (10 kg TRU/year) in equilibrium.

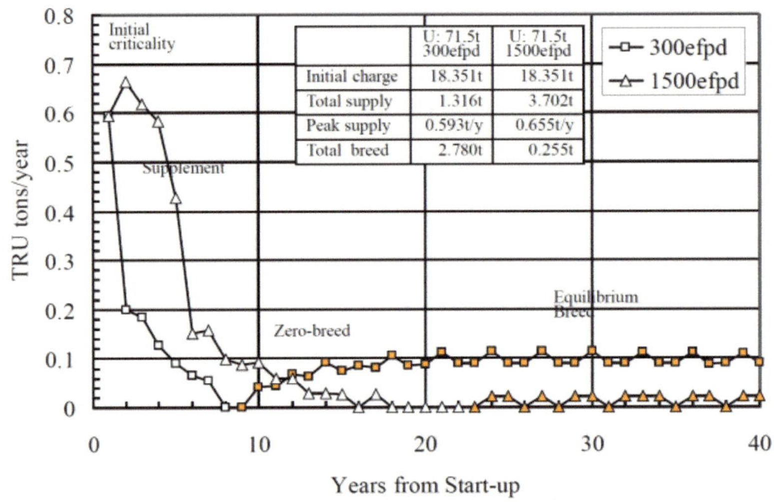

Figure 7.
Effect of fuel salt cleaning interval on the feed/breed balance.

2.6.3 Effect of initial fissile isotope composition

The effect of isotopic composition of initial feed TRU was evaluated for BWR-UOx fuel and ABWR-MOX fuel as shown in **Figure 8**. The isotopic compositions of each feed TRU are shown in **Table 8**.

It is revealed that the breeding performance of an FFMSR applied on the ABWR-MOX spent fuel is much better than that on the BWR-UOX spent fuel though they can be comparable after the equilibrium state.

What is more drastic is the capability of accumulated TRU to support deployment of the FFMSR. It is assumed that a 3.3 GWt (1.0 GWe) BWR yields annually 20.4 t of spent nuclear fuel (SNF) (50 GWd/t-U) containing 0.27 t of TRU; meanwhile a 3.93 GWt (1.38 GWe) full MOX ABWR yields annually 34.8 t of SNF (33 GWd/t-HM) containing 1.28 t of TRU. The accumulated SNF from a BWR for 54.6 years will support an FFMSR-UOX, and that of an ABWR's SNF for 17.8 years will support an FFMSR-MOX, with equivalent power output the same as the respective reactor. This means that a full MOX ABWR can be a breeding reactor with 17.8 years doubling time by the combination of FFMSR deployment.

2.7 Evolution of TRU constitution

The TRU inventory is almost kept at a constant through FFMSR operation with specific trends of isotopic evolution as shown in **Figures 9** and **10**.

The high content of Np is distinct in the TRU from LWR; however it is transmuted effectively. The content of Pu isotopes is getting saturated in both cases. Am isotopes are slowly decreasing until 40 years. The buildup of Cm is over a factor of 3.5; however it tends to be saturated after 20 years. This is a characteristic feature compared with the case of MOSART [29] in which non-fissionable Cm isotopes

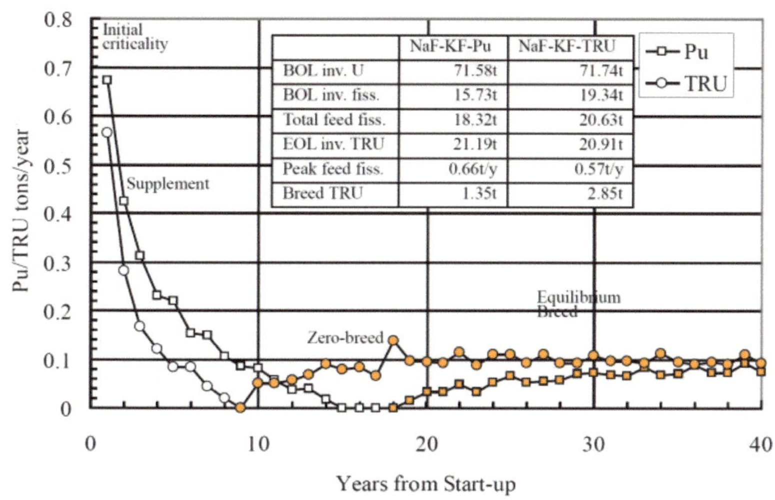

Figure 8.
Effect of initial fissile isotope composition on the feed/breed balance.

Source of TRU	Np/Pu/Am/Cm (wt.%)	$^{238/239/240/241/242}Pu$ (wt.%)
BWR-UOX-45GWd/t-U	5.19/89.22/4.90/0.69	2.80/51.77/25.98/11.07/8.38
ABWR-MOX-33GWd/t-HM	0.35/91.69/7.11/0.85	2.62/38.17/35.33/13.49/10.39

Table 8.
Isotopic composition of initial feed TRU.

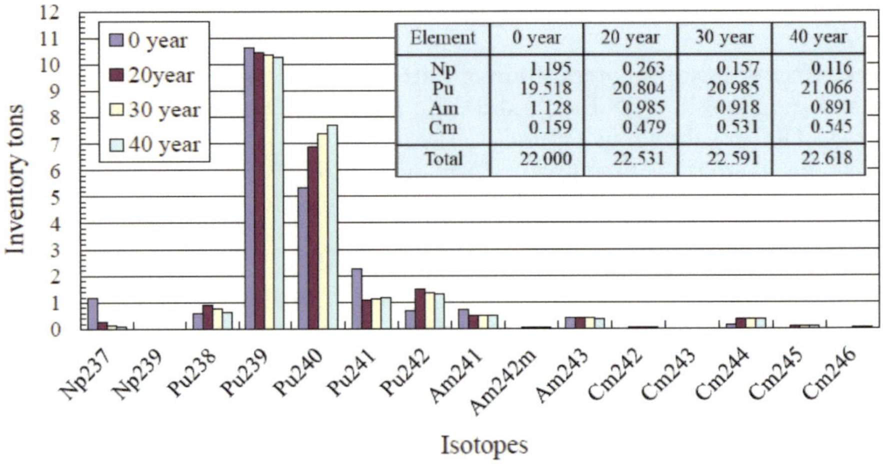

Figure 9.
Evolution of TRU isotopic composition during burnup (BWR-UOX).

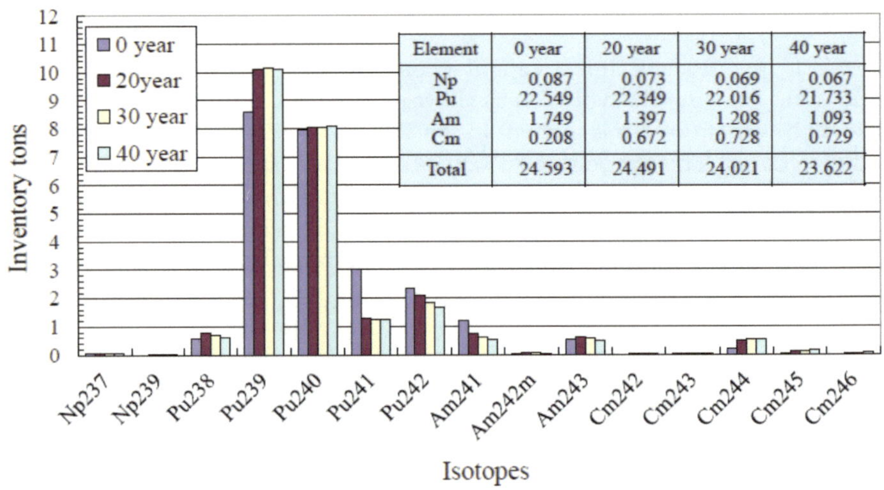

Figure 10.
Evolution of TRU isotopic composition during burnup (ABWR-MOX).

build up remarkably. Generally favorable features of fast neutron irradiation are represented, though further assessments for several hundred years are inevitable.

2.8 Freezing behavior of fuel salt

The molten salt reactor is feasible as long as the liquidus temperature of the fuel salt is kept at least 50°C lower than the reactor core inlet temperature. According to the classic design principle of molten salt reactors, the fuel salt should be composed of a single eutectic mixture, and all components of the fuel salt should congruously solidify at the eutectic point.

In the case of the FFMSR, the phase change is incongruous manner as the fuel salt should be composed of a pair of independent eutectic mixtures. It should be qualified by freezing behavior down to the solidus temperature in order to justify any engineering effort particular to the molten salt reactor such as the freeze valve, the fuel drain tank, and the reactor safety evaluation.

The freezing of NaF-KF-UF_4-PuF_3 system is dictated by the eutectic point of fissile salt (605°C) to give eutectic of NaF-KF-PuF_3 irrespective of the concentration of UF_4 as shown in **Table 9**. These values of liquidus temperature are substantially higher than that of the classic fuel salt such as 0.72LiF-0.16BeF_2-0.12ThF_4 (500°C) for thermal neutron molten salt reactors based upon the Hastelloy N technology however near to that of the revised MSFR (594°C) [30].

The solidified fuel salt eventually produces a specific stratified structure, a lighter fissile salt on a heavier fertile salt. The density of solidified salt is assumed as 8% higher than that of liquid at the same temperature.

Feasibility of the freeze valve can be controversial because it has originally been developed on the assumption that the fuel salt was a single eutectic mixture which solidified congruously.

2.9 Effect of burnup and tri-fluorides on freezing behavior

If the U(IV)/U(III) ratio in the system is fixed at 20 as a redox buffer medium, 71.4 tons-U (300,000 moles) of the total U inventory should consist of 285,700 moles of UF_4 and 14,300 moles of UF_3. The concentration of UF_3 is 1.33 mol% when that of PuF_3 is 8.10 mol%. Meanwhile, UF_4 inventory is reduced to a factor of 0.952 by chemical reduction to UF_3.

It has been suggested thermodynamically that tri-fluorides of fission product lanthanide behave as PuF_3 as well as those of minor actinide in the phase relationship and would interfere the freezing behavior.

Calculations are made to evaluate the effect of reduction of UF_4 to UF_3 and buildup of fission product lanthanide tri-fluorides in NaF-KF-0.281UF_4-0.081PuF_3 fuel salt according to chemical processing intervals for two cases of fissile salt arrangement and shown in **Tables 10** and **11**.

It is revealed that the effect of UF_4 reduction to UF_3 does not affect liquidus temperature of fuel salt meaningfully irrespective of fissile salt.

Temp. °C	Liquid Composition mole ratio	Wt. %	Density g/cc	Solid Composition mole ratio	Wt. %	Density g/cc
~605	0.354NaF-0.290KF-0.275UF_4-0.081PuF_3	100.00	4.700			
605	0.341NaF-0.146KF-0.275UF_4	76.81	4.585	0.013NaF-0.144KF-0.081PuF_3	23.19	5.443
580~	0.341NaF-0.146KF-0.189UF_4	57.80	4.156	0.086UF_4	19.00	6.72*
490				0.341NaF-0.146KF-0.189UF_4	57.81	4.595
~605	0.349NaF-0.289KF-0.281UF_4-0.081$TRUF_3$	100.00	4.727			
605	0.336NaF-0.145KF-0.281UF_4	77.95	4.624	0.013NaF-0.144KF-0.081$TRUF_3$	22.93	5.580
590~	0.336NaF-0.145KF-0.173UF_4	53.47	4.049	0.108UF_4	23.60	6.72*
490				0.336NaF-0.145KF-0.173UF_4	53.47	4.491

* at 25°C

Table 9.
Liquid and solid components of fuel salt during freezing.

Chem. Process Interval (EFPD)	UF_3 mole %	Fuel Salt (NaF-KF)-UF_4-$TRUF_3$					Fertile Salt (NaF-KF)-UF_4					Fissile Salt (NaF-KF)-$TRUF_3$				
		Composition (mole%)				liq.[a] °C	mole %	Composition (mole%)			liq.[a] °C	mole %	Composition (mole%)			liq.[a] °C
		NaF	KF	UF_4	XF_3[b]			NaF	KF	UF_4			NaF	KF	XF_3[b]	
0	0	34.62	28.88	28.10	8.10	605	76.18	44.18	18.94	36.88	590	23.82	5.3	60.7	34.0	605
0	1.33	33.33	30.49	26.75	9.43		72.26	44.09	18.89	37.02	600	27.74				
300		33.12	30.59	26.75	9.54		71.94	43.97	18.85	37.18	600	28.06				
600		32.91	30.69	26.75	9.65		71.62	43.86	18.80	37.34	605	28.38				
900		32.68	30.80	26.75	9.77	610	71.26	43.73	18.74	37.53	610	28.74				
1200		32.47	30.90	26.75	9.88	620	70.94	43.60	18.69	37.71	620	29.06				
1500		32.26	31.00	26.75	9.99	625	70.62	43.48	18.64	37.88	625	29.38				
1500	0	33.85	29.39	28.10	8.66	620	74.53	43.62	18.70	37.68	620	25.47				

liq[a], liquidus temperature; XF_3[b]: PuF_3 + UF_3 + LaF_3 + if any.

Table 10.
Option (a): to keep eutectic freezing at 605°C of fuel salt, 0.053NaF-0.607KF-0.340PuF_3.

Chem. Process Interval (EFPD)	UF_3 mole %	Fuel Salt (NaF-KF)-UF_4-$TRUF_3$					Fertile Salt (NaF-KF)-UF_4					Fissile Salt (NaF-KF)-$TRUF_3$				
		Composition (mole%)				liq.[a] °C	mole %	Composition (mole%)			liq.[a] °C	mole %	Composition (mole%)			liq.[a] °C
		NaF	KF	UF_4	XF_3[b]			NaF	KF	UF_4			NaF	KF	XF_3[b]	
0	0	35.29	28,51	28.10	8.10	610	76.79	44.39	19.02	36.59	580	23.21	5.2	59.9	34.9	610
0	1.33	33.76	30.06	26.75	9.43		72.98	44.34	19.00	36.66	590	27.02				
300		33.56	30.15	26.75	9.54		72.66	44.23	18.96	36.81	590	27.34				
600		33.36	30.24	26.75	9.65		72.35	44.12	18.91	36.97	590	27.65				
900		33.14	30.35	26.75	9.77		72.01	44.00	18.86	37.14	600	27.99				
1200		32.93	30.44	26.75	9.88		71.69	43.88	18.81	37.31	605	28.31				
1500		32.72	30.54	26.75	9.99		71.38	43.76	18.76	37.48	610	28.62				
1500	0	34.25	28.99	28.10	8.66		75.19	43.84	18.79	37.37	605	24.81				

liq[a], liquidus temperature; XF_3[b]: PuF_3 + UF_3 + LaF_3 + if any.

Table 11.
Option (b): to allow liquidus at 610°C of fuel salt, 0.052NaF-0.599KF-0.349PuF_3.

The buildup of lanthanide tri-fluorides does affect the liquidus temperature of fertile salt up to 625°C for the case (a); meanwhile it does not exceed 610°C using the fissile salt (b).

Option (a) should allow 900 EFPD of the chemical process interval if the liquidus temperature of fertile salt at 610°C is acceptable.

Option (b) should allow 1500 EFPD of the chemical process interval if the liquidus temperature of fertile salt at 610°C is acceptable. Option (b) however is against the rule in which no free fissile material is deposited before eutectic freezing. The choice of alternatives is depending upon less than 3% of difference of designated molar composition of tri-fluoride in the fissile salt. Not only the phase behavior of stable tri-fluoride such as PuF_3 and LnF_3 but also that of fluctuated UF_3 should be examined carefully.

3. Chemical processing

3.1 How fission product stream be free from TRU

It has been evaluated that the radiotoxicity of the PWR-UOX-SNF of 50GWd/t-U decreases to the reference level represented by that of annually transmuted natural uranium (7.83 t-$U_{nat.}$) after 130,000 years from discharge. If the HLW contains absolutely no TRU, the radiotoxicity decreases to the reference after 270 years mainly dominated by that of alkali and alkali earth elements (FP_{alk}: Rb, Cs, Sr., Ba) as shown in **Figure 11** [31].

The radiotoxicity of HLW from a reprocessing of UOX fuel with a nominal Pu loss rate of 0.5% and with removing minor actinides (MA; viz., Am and Cm) with a loss rate of 1% will decrease at the reference level in 500 years. It is assumed that the period will decrease to 370 years if Pu and MA are removed simultaneously from the HLW as TRU at the overall loss rate of 0.5%. This represents that the permissible TRU content in the finally disposed fission product (FP_{alk}: Rb, Cs, Sr., Ba) is 65.9 g-TRU/8461 g-FP_{alk} (0.78%) as shown in **Table 12.**

The nuclear fuel of a 3.2 GWt FFMSR supported by 93.6 t-HM reaches the burnup of 51.3 GWd/t-HM in 1500 EFPD by consuming depleted uranium (4.33 t-$U_{dep.}$/50 GWd/t-HM), which might have been discarded as a radioactive waste somehow. If the radiotoxicity of 4.33 t-U instead that of 7.83 t-U is assumed as the reference for the HLW of FFMSR, the period to decrease to the revised reference value might be extended to 500 years after discharge. In order to keep TRU/FP_{alk} at 0.78%, the permissible loss rate of the TRU into the FP_{alk} should be less than 0.036% due to the specific TRU concentration in an FFMSR fuel as high as in an equivalent LMFBR fuel, as shown in **Table 11.** The required loss rate is far less than

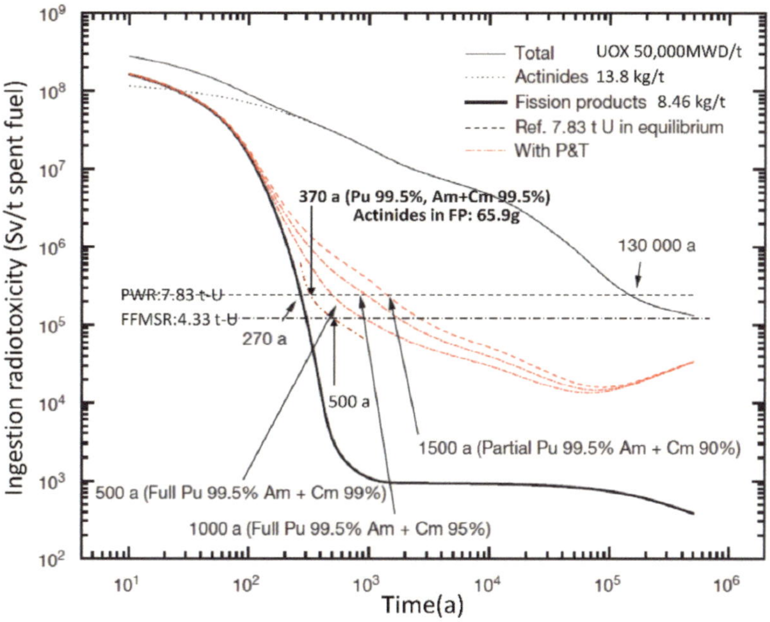

Figure 11.
Ingestion radiotoxicity of 1 t of spent nuclear fuel [31].

	PWR-UOX 50GWd/t-U			FFMSR 51.3 GWd/t-HM*		
Element	**Mass g/t-U**	**Permissible TRU (g)**	**Required loss rate**	**Mass g/t-U**	**Permissible TRU (g)**	**Required loss rate**
Uranium	935,245			730,300		
TRU	13,179			212,100		
Halogens	358			741		
Rare gases	8388			7533		
Noble and semi-noble metals	13,306			18,594		
Alkali and alkaline earths	8461	65.9	0.5%	9665	75.3	0.036%
Lanthanides	15,621			16,655		
Zirconium	5442			4559		
FP total	51,576			57,624		

**Burnup at the end of the first 1500 EFPD and thereafter.*

Table 12.
Comparison of required loss rate.

0.1% of the target to be achieved by the pyro-processes such as electrochemical refining or liquid metal extraction currently under development [31].

The chemical processing in the FFMSR should be efficient to remove fuel material from the fission product streams but not necessarily efficient to remove fission products to become as neutron poisons if it were operated under the thermal neutron from the fuel stream.

To perform this new and perpetual mission, a processing interval of 1500 EFPD is sufficiently long and provides a small throughput in other words. The online chemical processing facility of α-β-γ-n remote operable capability collocated with the FFMSR would be the most expensive auxiliary part of the plant to be constructed as well as to be operated. Such cost should be depending upon the nature of process, i.e., process complexity, material compatibility, process wastes, and capacity in particular.

3.2 Requirements to be concerned

The crucial point in the fuel cleanup process is not the complete removal of neutron-absorbing material such as lanthanides from the fuel but keeping any leak of actinides into waste streams as low as possible. This could justify the use of the selective oxide precipitation process as an absolutely simple choice compared with other pyro-processes such as the electrochemical or the reductive extraction [32].

The fluoride volatile process of UF_6 had been perceived as the most practical since the successful operation in MSRE during switch over the fissile from ^{235}U to ^{233}U; however it has been overlooked the fact that metallic Zr scrap in addition to the fuel should be followed by a prolonged H_2 sparge to remove metallic corrosion products (Ni, Fe, Cr) caused by F_2 treatment. The presence of a certain amount of Pu should require applying a reducing process from PuF_4 to PuF_3 in order to avoid accidental precipitation of PuO_2 and severe material corrosion. Any absence of such treatment after the final removal of $^{233}UF_6$ might have resulted MSRE remediation in a fruitless and endless trouble by undisclosed reasons of line clogging of the fuel drain tank.

3.3 Selective oxide precipitation process

In the very early stage of the Molten-Salt Reactor Program (MSR Program) started at ORNL, experimental studies on selective precipitation of oxides had been carried out because it might have been a suitable scheme for the reprocessing of molten salt reactor fuels, though it was abandoned after the discovery of the reductive extraction and metal transfer process associated with the UF_6 volatile process, which, though complex and material incompatible, involved handling only liquids and gases. However the ultimately small throughput may allow us to select a solid handling process if the process is simple, fast, and material compatible.

A successful attempt was made to precipitate mixed uranium, plutonium, minor actinides, and rare earths from LiF-NaF molten salt solution by fluor-oxide exchange with other oxides (e.g., CaO, Al_2O_3) at temperatures 700–800°C. It was found that the following order of precipitation in the system is U-Pu-Am-Ln-Ca. Essentially all U and TRU were recovered from the molten salt till to rest concentration 5×10^{-4}%, when 5–10 mol% of rare earths are still concentrated in solution [33, 34].

An optional process to be applied to the DMSR fuel was suggested as follows. Treat the melt with a strong oxidant to convert UF_3 to UF_4, PaF_4 to PaF_5, and PuF_3 to PuF_4. Precipitate the insoluble oxides using water vapor diluted in helium. The oxides UO_2, Pa_2O_5, PuO_2, CeO_2, probably NpO_2, and possibly AmO_2 and CmO_2 should be obtained. Recover the oxides by decantation and filtration. Hydrofluorinate the oxides into the purified melt of LiF-BeF_2-ThF_4, and reduce the melt with H_2 and reconstitute fuel with the desired UF_4/UF_3 ratio [35].

This could justify the use of the selective oxide precipitation process as an absolutely simple choice compared with other pyro-processes such as the electrochemical or the reductive extraction [36].

3.4 Customization of the process

Based upon the survey, it is concluded that the application of the selective oxide precipitation process with alkali or alkali earth metal oxides (K_2O_2; melt at 490°C and CaO; solid) as the oxidizer can be feasible under special cautions about selectivity to the FFMSR technology relying on NaF and KF as major constitutes of fuel solvent, as shown in **Figure 12**.

If it can reduce TRU concentration to 5×10^{-4} mol% in liquid phase from 8 mol %, the available loss rate will be 6.25×10^{-5}. The permissible loss rate of 3.6×10^{-4} is six times larger than the available loss rate.

Intense increase of liquidus temperature should be taken into account during actinide removal treatment from 605°C up to 800°C. Using K_2O_2 as a precipitator can modify Na/K ratio from 0.55/0.45 to nearly 0.40/0.60 to give eutectic mixture at 710°C.

Elemental fluorine freed from UO_2 precipitation reaction would react with $TRUF_3$ to oxidize them into $TRUF_4$ which can be eventually precipitated as $TRUO_2$ by succeeding the use of CaO as a precipitator no more than ca. 20 mol% which may give stable ternary eutectic at ca. 700°C of the final waste salt.

As actinides are extremely abundant than lanthanides, the separation efficiency of actinides from lanthanides should not be good enough in a practical application; repeated treatments might be required to reduce actinide concentration in the lanthanide stream until permissible level is attained, even though moderate amount of lanthanides are permitted in the actinide stream. Up to 10% of lanthanides would be allowed to leave in the fuel salt stream, but lower than 0.01% of actinide leak into the waste stream is anticipated.

The process is a small batch scale (e.g., 21.2 l/day) in a pure Ni-made vessel facilitated to eliminate solid handling but performed by liquid phase handling only. The relevant fuel batch contains 12.9 kg of TRU which substantially exceed the significant mass of 8 kg; however it is always accompanied with 47 kg of chemically inseparable uranium. It is anticipated that the heat generation rate of a fuel batch will be 13.4 kW and the radioactivity will be 6MCi at 2 days after being drained.

The process is incorporated with He sparge to purge rare gases and halogens as well as noble and semi-noble metal fission products and electroreductive removal of zirconium developed for the MSRE remediation [37] as shown in **Figure 13**. Accumulation of fission product zirconium tetrafluoride in the fuel system would

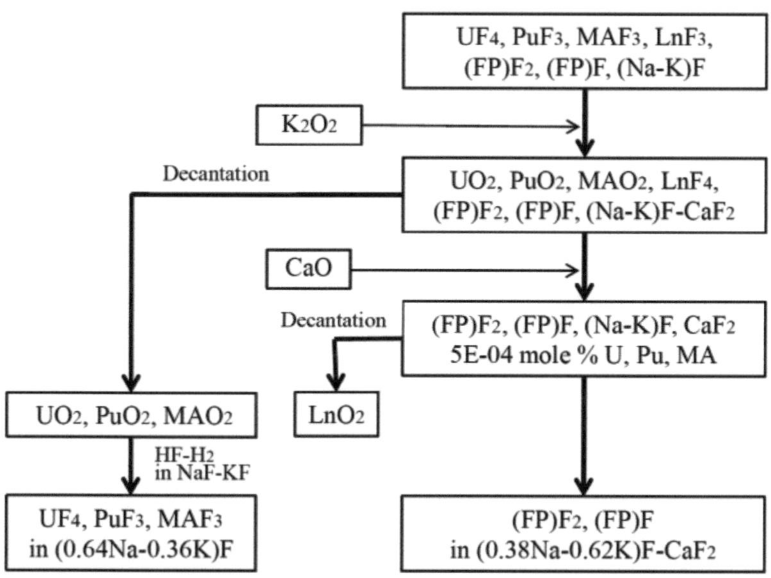

Figure 12.
Process flowsheet of the oxide selective precipitation.

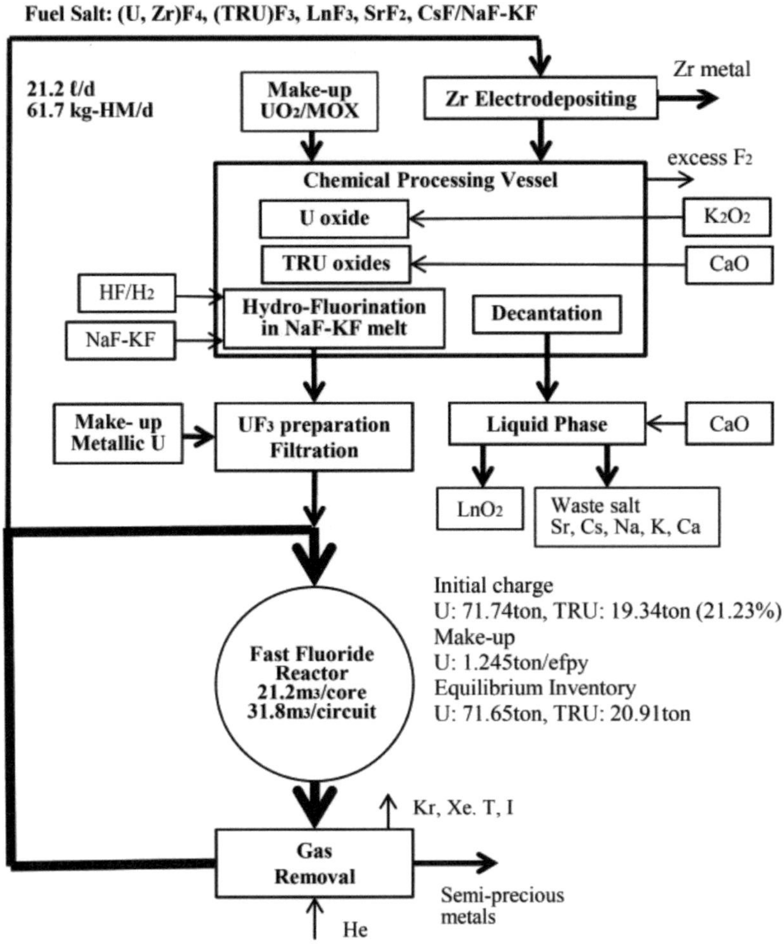

Figure 13.
Online chemical process in a typical 3.2 GWt FFMSR.

Fuel-in mole%		K_2O_2 26.52 mole%	CaO 19.14 mole%	Decant -ation	Hydro-fluorination in NaF-KF	Fuel-out U make-up		CaO 2.08 mole%	Waste solution mole%	
UF_4	26.67	UO_2	UO_2		UF_4	UF_4	27.27			
UF_3	1.33					UF_3	1.40			
NpF_3	0.03	NpF_4	TRU-O_2		NpF_3	NpF_3	0.03			
PuF_3	7.63	PuF_4			PuF_3	PuF_3	7.63			
AmF_3	0.26	AmF_4			AmF_3	AmF_3	0.26			
CmF_3	0.17	CmF_4			CmF_3	CmF_3	0.17			
LnF_3	1.04	LnF_4		LnF_4				LnO_2		
RbF CsF SrF_2 BaF_2	0.31	RbF CsF SrF_2 BaF_2		RbF CsF SrF_2 BaF_2 CaF_2					RbF CsF SrF_2 BaF_2 CaF_2	15.65
NaF	43.79	NaF					44.27		NaF	32.05
KF	18.77	KF					18.97		KF	52.30

Table 13.
Process parameters of oxide selective precipitation.

give an adverse effect in the fuel storage tank due to its reducible nature under gamma radiation as well as sublimation. Some detail process parameters are shown in **Table 13**.

4. Chemical engineering of FFMSR

4.1 Initial fuel charge

An institutional restriction imposed to our task is the fact that no separated plutonium is tolerable in Japan to secure proliferation resistance under the international agreement. Japanese reprocessing plant cannot produce anything but U-Pu mixed oxide.

In the case of the FFMSR, the preparation work of initial charge does not require a high gamma facility if the source materials come from a conventional reprocessing plant. The oxide precipitation process incorporated with the hydro-fluorination process makes solid mixed oxide as makeup material feasible.

A typical 3.2 GWt FFMSR requires U-21.23% TRU mixed compound of 90 tons for the initial charge and 3.41 kg-U/EFPD (1245 kg-U/EFPY) of makeup in the equilibrium state compared with the 47.8 kg-U/EFPD of projected throughput of the chemical processing.

The FFMSR requires several tons of TRU supplement according to the nuclear characteristics until it reaches to equilibrium. This system is capable of making up 0.92 kg-TRU/EFPD (336 kg-TRU/EFPY), if the same U-TRU mixed compound as the initial charge is applied.

According to the specific nucleonic characteristics, the minimum U makeup is 1115 kg-U/EFPY, and the peak TRU supplement is 720 kg-TRU/EFPY. This means that as high as 39.2% U-TRU mixed compound should be temporally required in this occasion.

4.2 Redox buffer control and burnup effect

The nuclear reaction in the FFMSR consists of transformation of UF_4 into $TRUF_3$ and fission of $TRUF_3$ into fission products. The annual free fluorine production of 3.2 GW_{th} FFMSR at the equilibrium is 1308 moles (0.25/0.238 mol/kg-U $\times$ 1245 kg-U/EFPY) from the transmutation of UF_4 and 3264 moles from the fission of $TRUF_3$ based on 1.02 mole-F/MWt/y times 3200 according to **Table 6**. The annual consumption of UF_3 is 4572 moles (1088 kg-U). This can be compensated by dissolution of 1524 moles uranium metal (363 kg-U) in the fuel salt containing UF_4 as a part of annual U makeup (1245 kg-U), though any side stream hydro-fluorination is also available.

Taking into account uranium inventory as much as 71.65 tons (28 mol%), assumed U[IV]/U[III] = 20 ratio represents 3.41 tons of U[III] inventory and 1.33 mol% of UF_3 concentration. Since the daily supply of U[III] is 3 kg/EFPD, very stable control of U[IV]/U[III] ratio is available. On the other hand, steadiness of UF_3 concentration as high as 1.33 mol% represents 26.67 mol% of the UF_4 and 9.33 mol% of the total tri-fluoride concentration instead of 8.0 mol% of $TRUF_3$.

It should be assumed that the inventory of fission product lanthanide tri-fluoride at the burnup of 50,000 MWd/t-HM is 6.9% (0.55 mol%) of TRU tri-fluorides. Any effect of fluctuation as high as $\sim$1.33 mol% in UF_4 or $\sim$1.88 mol% in total tri-fluoride upon the liquidus temperature of fuel salt should be carefully examined.

4.3 Back-end process and radioactive wastes

In the FFMSR, the inventory ratio of fission products to that of TRU is the key factor to guarantee an effectively low concentration of TRU in the waste stream with a given TRU leak rate. The inventory of fission product is equal to that of accumulation during 1500 EFPD (51GWd/t-HM).

The waste stream consists of gases (He, Kr, Xe, and ^{3}H), spent charcoal filter absorbing I, solid elements (Zr, rare metals, and semi-rare metals such as Zn, Ga, Ge, As, Se, Nb, Mo, Ru, Rh, Pd, Ag, Tc, Cd, In, Sn, Sb, and Te), lanthanide oxides, and NaF-KF-CaF_2 matrix salt containing alkali/alkali earth fission product fluorides.

Storage of fission product gases in high-pressure cylinders and then transfer to the repository was a standard practice in the MSBR design; however it is impractical in Japan, because of regulative requirement of annual pressure proof test of high-pressure cylinders.

Though the fission yield of ^{85}Kr from the FFMSR system is assumed as about 1/3 of that from the graphite-moderated thorium molten salt reactors, special attention was suggested such as underground disposal by geological hydro-fracturing should be paid for radioactive Kr [38] if releasing from a high stack as currently applied in the spent fuel reprocessing plant will not be allowed in the future.

The spent iodine filter such as silver-impregnated matrix is a universal issue in every molten salt reactor as well as in spent solid fuel reprocessing plants.

Zr is electrochemically separated from the fuel salt prior to the oxide precipitation. Zr compounds are not desirable in the waste salt tank because of their reducibility in addition to sublimation capability [34].

Rare and semi-rare metals could possibly be industrially utilized after appropriately separated because they are virtually alpha activity free. They include various very long-lived fission products, such as ^{99}Tc, ^{126}Sn, ^{79}Se, and ^{107}Pd, which are to be disposed in a very compact form.

NaF-KF mixture containing soluble and major heat-generating fission product fluorides (CsF, SrF_2, etc.) and the process reagent (KF and CaF_2) is the main process waste as far as the online chemical processing is concerned.

Composition of fuel salt is assumed as 0.348NaF-0.284KF-0.280UF_4–0.082$TRUF_3$-0.006LnF_3 ($T_{liq.}$ = 605°C), and that of waste salt is assumed as 0.356NaF-0.580KF-0.060CaF_2-0.004$FPF_{1.5}$ (T_{liq} = 700°C).

Storage of the waste salt as liquid phase at higher than 700°C should be unpractical. It might be cooled to solidify in a tank shortly after being transferred.

The inventories are assumed as fuel salt, 147.87 tons; HM fluoride, 120.54 tons; and matrix salt, 27.33 tons. The high-level waste salt originated from a 1.5 GWe FFMSR system for 1500 EFPD operation (51.3 GWd/t-HM) is 46.26 tons (20.12 m^3 at 2.3 g/cc of density), and the radiotoxicity of this amount of waste is equivalent to 405 tons of depleted uranium after 500 years cooling.

The throughput of high-level waste salt mixture from the vitrified high-level waste of 1.5 GWe PWR (50 GWd/t-U) after 99.5% Pu by reprocessing and 99% MA removal by P&T is probably 59 tons, and the radiotoxicity of this amount of waste is equivalent to 1163 tons of natural uranium after 500 years cooling.

The selection of the fuel matrix without ^{7}Li economically allows a direct disposal of the waste matrix salt without recycle; nevertheless the bulk mass is comparable to that of vitrified waste of LWR though public utilization of decay heat before immobilization of cooled waste salt might be feasible.

Furthermore the incomparably favorable fact that the FFMSR system does not produce any fuel cycle-associated wastes, starting from uranium mine tailing all through to alpha-contaminated HEPA filters of MOX fuel fabrication plant, should be taken into account.

The characteristic capability of the oxide selective separation process enables to retrieve alpha contamination-free metals as well as lanthanide oxides without elaborating partitioning processes. Effective technologies to utilize such recovered resources are sincerely expected.

Full deployment of the FFMSR should make the entire fuel cycle infrastructures from the uranium mining to the spent fuel reprocessing including P&T needless except the HLW disposal site.

4.4 Contingency plan

The annual loss of TRU due to fuel salt chemical cleaning is 6 kg based upon the assumption 1500 EFPD of interval and 0.1% of nominal loss rate for 22.6 tons-TRU inventory. This can be accounted for in the equilibrium phase indefinitely because the annual TRU surplus is 10 kg. However if a flushing procedure should be required at the maintenance work according to 0.43% of the transfer rate in the MSRE operation experience [34], 97 kg of TRU may be transferred to the flushing salt even if it will be recovered efficiently later. How much TRU should have been given as a dowry at the deployment of a stand-alone FFMSR is a question. The reactivity swing by the chemical process unit outage (halt of the makeup and FP separation) should also be evaluated.

4.5 Dedicated front-end process for the ABWR

The dedicated front-end plant might produce U-TRU mixed fluoride from the MOX spent fuel of ABWR for which the Rokkasho Reprocessing Plant cannot deal with technical reasons as shown in **Figure 14**.

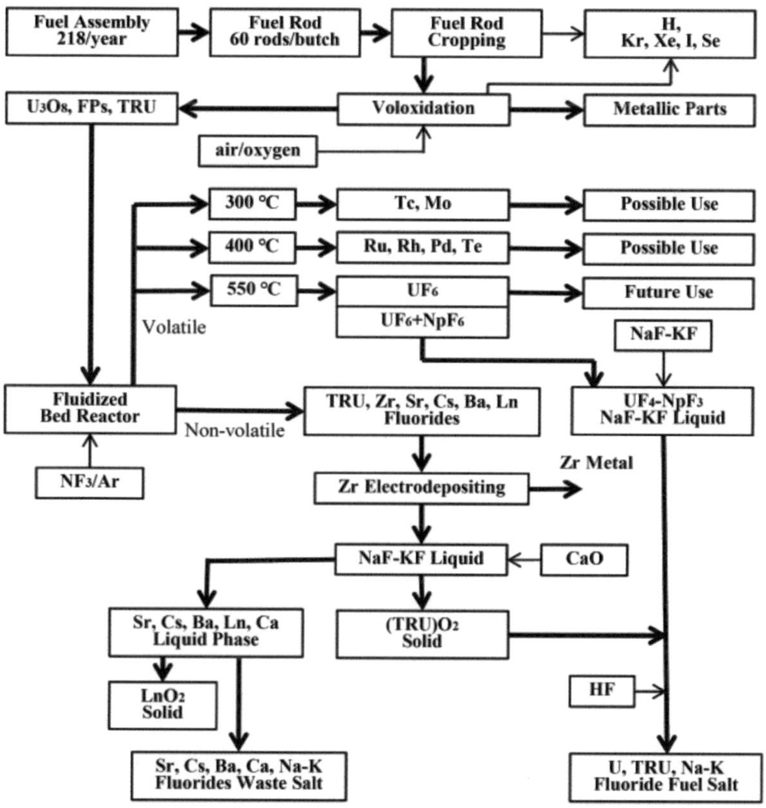

Figure 14.
Dedicated front-end process for the ABWR-MOX fuel.

The original fluoride volatility process converts all components into volatile fluorides by using fluorine flame reactor and then separates them into fractions according to their properties [39]. However we were rather interested in the recently developed innovative process using NF_3 as a thermally sensitive reagent; it would react with different compounds at different temperatures [40]. For example, NF_3 reacts with Tc and Mo oxide near 300°C and Ru and Rh near 400°C, while U oxides required near 500°C to form a volatile fluoride. This process eventually yields the nonvolatile fraction containing all TRU fluorides. Then we intended to apply the oxide selective precipitation process, to provide TRU stream not so much cleaned from fission products but to result very clean fission product stream from TRU contamination.

The distinguished feature of this process is the capability to separate useful metallic fission products as well as lanthanide oxides free from alpha contamination from other residual materials of fluorination process effectively, without laborious partitioning.

A suite of processes are shown as the flowsheet specifically for the ABWR spent fuel processing; however it can be reasonably modified to the original LWR spent fuel or LWR-MOX spent fuel.

5. Experimental test plans

5.1 Clarify phase relationship in NaF-KF-UF_4-UF_3-PuF_3 system for the FFMSR

It is perceived that experimental confirmation of density assessment procedure of molten salt mixtures is inevitable to establish any MSR technology. The liquid fuel of the FFMSR contains UF_4, UF_3, and PuF_3. Currently any performance of experimental activity on the specimens containing Pu as the special nuclear material is not available other than in the Russian Research Laboratories.

We plan the experimental procedure using NaF-KF-$^{nat.}UF_4$ containing in situ prepared $^{nat.}UF_3$ to simulate NaF-KF-$^{nat.}UF_4$-PuF_3 taking advantage of identical crystal structure as well as similarity of density between PuF_3 and UF_3.

Furthermore, the phase relationship (freezing behavior) will be experimentally evaluated in order to justify that the feasibility of the phase structure should be understood.

The plan includes:

1. Confirmation of synthetic process of heavy element fluoride.
2. Confirmation of recovery process of heavy element as UO_2.
3. Confirmation of synthetic process of NaF-KF-UF_4-UF_3.
4. Density measurement of liquid NaF-KF-UF_4-UF_3 to clarify the dependency of heavy element content with different solid densities on density of the liquefied salt.
5. Investigation of the phase diagrams of NaF-KF-UF_4-UF_3 to clarify the dependency of UF_3 collocation in the NaF-KF-UF_4 phase diagram using the solubility measuring practice. Effect of trivalent fission products on the phase diagram using CeF_3 as a surrogate of UF_3 and PuF_3.

5.2 Experimental confirmation of chemical effects of TRU fissioning

The chemical effects of UF_4 fissioning in a fluoride molten salt reactor were confirmed by the successful operation of the MSRE during the end of the 1960s. However any experimental confirmation of the chemical effect of PuF_3 fissioning in a fluoride molten salt reactor has not yet been undertaken in spite of a strong warning made by the ORNL scientist in the end of the last century [23].

In spite of the continued effort by the author to try to stimulate academic discussion on the chemical effect of TRU fissioning controversial against the ORNL since 2015, it seems to the author to become "an inconvenient truth" for which no one dares to discuss. The author seriously concerns that the present situation might jeopardize the technological development of plutonium burning technology in the immediate future.

The author plans to propose a capsule irradiation test of NaF-KF-$TRUF_3$ specimens under the fast neutron flux (3.9×10^{19} m^{-1} s^{-1}) during liquid Na cooling in an experimental fast reactor (JYOYO) located in Oharai, Japan. It plans to measure the freed fluorine ions per a fission of fissile Pu and compare with that of ^{235}U by the weight loss of the pure Zirconium metal specimen immersed in the fuel salt.

The proposed specimens are:

1. 0.053NaF-0.608KF-0.340$TRUF_3$ eutectic mixture (liquidus: 605°C) 2.56 g-TRU/cc as the subject.

2. 0.053NaF-0.608KF-0.340CeF_3 eutectic mixture (liquidus: 605°C) as the reference.

3. 0.528NaF-0.285KF-0.188$^{235}UF_4$ eutectic mixture (liquidus: 490°C) 2.52 g-U/cc as the comparative.

The nominal sample temperature in the test region is at least 600°C; however it is assumed that the gamma heat of capsule structure should enable to heat the specimen up to 750°C.

6. Conclusions

The study on our FFMSR was started from the review of the reference technology and based upon the comprehension of immaturity of the TRU burning technologies using the MSR due to the prejudice of the original design principle of ORNL in which the use of PuF_3 had been an exclusively temporary issue.

The various aspects but restricted in chemical technology discussed in this work should be taken into account and reviewed carefully in the imminent future activity although they are in limited scope and hypothetical nature to be verified experimentally. The present neutron physical calculations are preliminary nature in which the direct fission fraction of ^{238}U is not quantified, taking for instances. The system has not yet been optimized, in various factors.

FFMSR should provide us with a tool to stimulate immediate use of existing LWR by making values to the spent fuel as well as to the depleted uranium and to create nuclear fission energy not relying on the existing fuel cycle infrastructure with the ultimate safety owing to the absence of and eliminating fuel cycle wastes and the simplicity for an indefinitely long term.

One of a price in return for these efforts is exclusive challenges to overcome increased reactor core inlet temperature up to 660°C (50°C higher than the liquidus temperature of fuel) however it might deserve.

Acknowledgements

The author would like to express his thanks to Prof. Dr. Koshi Mitachi and Prof. Dr. Yoichiro Shimazu for their volunteer dedication to the entire physical calculations in this work, Prof. Dr. L.I. Ponomarev for the courtesy of personally introducing his work, and Standard Power, Co. Ltd. for publishing this chapter. He would like to dedicate this paper to the late Prof. Dr. Yoich Takashima for his guidance in initiating the work associated with MSR technology.

Author details

Yasuo Hirose
Standard Power, Inc., Tokyo, Japan

*Address all correspondence to: yahirose@mint.ocn.ne.jp

References

[1] Mourogov A, Bokov P. Potentialities of the fast spectrum molten salt reactor concept: REBUS-3700. Energy Conversion and Management. 2006;**47**: 2761-2771

[2] Benes O, Cabet C, Deloech S, Hosnedl P, Ignatiev V, Kornings R, et al. Review report on liquid salts for various applications. In: ALISIA Project. 2009. p. 15/64

[3] Lizin A, Tomilin S, Gnevashov O, Gazizov A, Osipenko A, Kormilitsin V. PuF_3, AmF_3, CeF_3, NdF_3 solubility in LiF_NaF_KF melt. Atomic Energy. 2013; **115**(1):11-17

[4] Lizin A, Tomilin S, Gnevashov O, Gazizov A, Osipenko A, Kormilitsin M. UF_4 and ThF_4 solubility in LiF_NaF_KF melt. Atomic Energy. 2013;**115**(1):22-25

[5] Lizin A, Tomilin S, Ignatiev V. Joint solubility of CeF_3 and PuF_3 in ternary melts of lithium, thorium and uranium fluorides. Radiochemistry. 2015;**57**(1): 32-37

[6] Lizin A, Tomilin S, Naumov S, Ignatiev V, Nezgovorov N, Baranov A. The study of joint solubility of UF_4 and PuF_3 in molten fluorides of lithium, sodium and potassium. Radiochemistry. 2015;**57**(5):425-429

[7] Degtyarev A, Pnomarev L. Molten salt fast reactor with U-Pu fuel cycle. Progress in Nuclear Energy. 2015;**82**: 33-36

[8] Hirose Y, Mitachi K, Shimazu Y. Feasibility of molten salt fast reactor for emerging national tasks. In: Proc. 2015 ANS International High-Level Radioactive Waste Management (IHLRWM 2015), Charleston, SC; April 12–16; ANS. 2015. pp. 608-617

[9] Hirose Y, Mitachi K, Shimazu Y. Operation control of molten salt U-Pu fast breeder reactor. In: Proc. ICAPP 2016, San Francisco, CA, April 17-20. 2016. p. 16124

[10] Hirose Y. Feasibility of using (Li, Na, K)F-UF_4-$TRUF_3$ fuels for U-Pu fast-spectrum molten-salt reactors. In: Proceedings of ICAPP 2017, Kyoto, Japan, April 28. 2017. p. 17107

[11] Hirose Y. Some chemical aspects of molten-salt reactors specifically in TRU burning applications. In: Proc. ICAPP 2018, Charlotte, NC, US, April 8-11. 2018. pp. 787-798

[12] Ponomarev L, Lizin A, Tomilin S, Fedorov Y, Hirose Y. Fast spectrum, liquid fueled reactors. In: Dolan T, editor. Molten Salt Reactors and Thorium Energy. Cambridge, MA, USA: Wood Head Publishing of Elsevier; 2017. pp. 376-433

[13] Grimes W, Cuneo R, Blankenship F, Keilholtz G, Poppendick H, Robinson M. In: MacFerson H, editor. Fluid Fuel Reactors. Reading, Mass, USA: Addison-Wesley Pub. Co.; 1958. pp. 569-591

[14] Benes O. Thermodynamics of molten salts for nuclear applications [thesis]. In: JRC Technical Notes, JRC-ITU-TN-2008/40. Prague: Institute of Chemical Technology; 2008

[15] Thoma R, editor. Phase Diagrams of Nuclear Reactor Materials, ORNL-2548; 1959

[16] Benes O, Kornings R. Actinide burner fuel: Potential compositions based on the thermodynamic evaluation of MF–PuF_3 (M = Li, Na, K, Rb, Cs) and LaF_3-PuF_3 systems. Journal of Nuclear Materials. 2008;**377**:449-457

[17] Canter S. Density and viscosity of several fluoride mixtures. In: ORNL-TM-4308; 1973

[18] Grines W, Cuneo D. Molten salt as reactor fuels. In: Tipton C, editor. Reactor Handbook. Vol. I. New York: Wiley (Interscience); 1960. pp. 425-473. Chap. 17

[19] Thoma R. Chemical Feasibility of Fueling MSR with PuF3. In: ORNL-TM-2256; 1968

[20] Ponomarev L. Private communication. November 26 2015

[21] Forsberg C. Accident criticality safety for fast spectrum molten salt reactors. In: Transactions of the 2007 ANS Annual Meeting, June 24–28 2007

[22] Grimes W. Molten-salt reactor chemistry. Nuclear Applications & Technology. 1970;**8**:137-155

[23] Toth LM, Del Cul G, Dai S, MetCalf G. Molten fluoride fuel salt chemistry. In: International Conference on ADTTA, Las Vegas, NV, AIP Conference Proceedings 346. 1994. pp. 617-619

[24] Holcomb D. Overview of fuel and coolant salt chemistry and thermal hydraulics. In: Presentation for US Nuclear Regulatory Commission Staff, November 7-8, 2017, Washington, DC. 2017. ML17331B115

[25] Shimazu Y, Hirose Y. Evaluation of redox environment in fluoride molten salt reactors based on ENDF-VII. In: Proceedings of AESJ 2015 Autumn Meeting, Shizuoka, Japan, September 10–12. 2015. B-20

[26] Baes C Jr. The chemistry and thermodynamics of molten salt reactor fuels. Journal of Nuclear Materials. 1974; **51**:149-162

[27] Ignatiev V, Feynberg OI, Gnidoi I, Merzlyakov A, Surenkov A, Uglov V, et al. Molten salt actinide recycler and transforming system without and with Th–U support: Fuel cycle flexibility and key material properties. Annals of Nuclear Energy. 2014;**64**:408-420

[28] Williams D. REDOX control of MSR fuel. In: GEDEON-Practis Workshop, Cadarache, France, June 19–20. 2002

[29] Ignatiev V, Feynberg O, Gnidoi I, Merzlyakov A, Smirnov V, Surenkov A, et al. Progress in development of Li,Be, Na/F molten salt actinide recycler & transmuter concept. In: Proceedings of ICAPP 2007, Nice, France, May 13–18. 2007. p. 7548

[30] EVOL (Project no.249696) Final Report. Available from: http://cordis.europa.eu/docs/results/249/249696/final1-final-report-f.pdf#search=%27EVOL%28Project+no.249696%29 Final+Report%27

[31] IAEA. Implications of partitioning and transmutation in radioactive waste management. In: Technical Reports Series No. 435. 2004. p. 5

[32] Shaffer J. Chemical reactions of oxides with fluorides in LiF-KF. In: ORNL-2474. 1958. pp. 99-101

[33] Shaffer J, Watson G, Grime W. Chemical reprocessing of reactor fuels by oxide precipitations. In: ORNL-2931. 1960. pp. 84-90

[34] Gorbunov V. Experimental studies on interaction of plutonium, uranium and rare earth fluorides with some metal oxides in molten fluoride mixtures. Radiochimija. 1976;**17**:109-224

[35] Engel J, Grimes W, Bauman H, McCoy E, Dering J, Rhoades W. Conceptual design characteristics of a denatured molten-salt reactor with once-through fueling. In: ORNL-TM-7207. 1980

[36] Ignatiev V, Gorbunov V, Zakirov R. Fuels and fission products clean up for molten salt reactor of the incinerator type. In: ISTC Task #1606. 2007

[37] Peretz J. Identification and evaluation of alternatives for the disposition of fluoride fuel and flush salts from the MSRE at ORNL. In: ORNL-ER-380. 1996

[38] Messenger S, Forsberg C, Massie M. Gaseous fission product management for molten salt reactors and vented fuel system. In: ICAPP-2012, Chicago, USA, June 24–28. 2012. p. 12097

[39] Uhlir J, Marecek M. Fluoride volatility method for reprocessing of LWR and FR fuels. Journal of Fluorine Chemistry. 2009;**130**:89-93

[40] McNamara B, Casella A, Scheele R, Kozelisky A. Nitrogen trifluoride-based fluoride-volatility separations process: Initial studies. In: PNNL-20775. 2011

Section 2

Nuclear Power Plant

Chapter 6

Calculation of the Dose for Public Individuals Due to a Severe Accident at the Angra 2 Nuclear Plant, Brazil

André Silva de Aguiar, Seung Min Lee and Gaianê Sabundjian

Abstract

Through a severe accident at nuclear power plant Angra 2, the whole body dose effective of the individuals members of the public located in the Emergency Planning Zones (EPZs) will be calculated, and later, the protective actions in these EPZs will be analyzed. Two different scenarios of radionuclide release into the atmosphere will be considered. In the first scenario, 2 h of the release of Xe, Cs, Ba, and Te, and the second scenario, 168 h of release.

Keywords: MELCOR, CALPUFF, atmospheric dispersion, nuclear power plant, public individuals dose

1. Introduction

From the nuclear accidents that occurred in the world [1–3], the International Atomic Energy Agency (IAEA), together with the licensing bodies of countries that use nuclear energy, requested that they carry out computer simulations of some accidents that are considered credible for their facilities, in order to verify their integrity when subjected to such events.

These accidents are known as the design basis, in other words, accidents of loss of primary coolant by large or small ruptures at points in the primary circuit, whose probability of occurrence is critical to the system. It was after the accident at Unit 2 of the Three Mile Island Nuclear Plant (TMI) in 1979 that it was necessary to study the more severe accidents.

These severe accidents are those in which substantial damage is expected in the reactor core [4]. The knowledge of accidents with core meltdown is based on the simulations with computer programs of the type MARCH [5], APRIL [6], MELCOR [7–9], SCADAP/RELAP5 [10–11], and MAAP4 [12]. Among these programs, MELCOR, SCDAP/RELAP5, and MAAP4 are widely used for the integral analysis of the melted material in the core and the consequences in the lower part of the pressure vessel [13].

In the event of a severe accident, followed by successive failures of physical barriers and problems in the control and protection systems of the reactor, the release of radioactive material into the atmosphere may become significant. The problems generated by these catastrophic events can lead to an increase in levels of radioactivity in the vicinity of the plant, representing a threat to the society and local life.

Therefore, the dispersion study can generate results with impacts on the occupation and dimensioning of the site. Also in this context, it is important to remember that the severity of a possible accident associated with nuclear facilities in general is strongly linked to population density of the regions around the facility, as well as, evacuation policy, medical treatment, and other health measures which should be taken to mitigate its radiological consequences.

2. Description of the accident

A typical PWR modeling was developed by the German company, GRS (Global Research for Safety), and supplied to CNEN, as shown in **Figure 1**. This modeling was chosen, in this study, for the purpose of performing an independent analysis of severe accidents in ANGRA 2 nuclear power plant (NPP). However, although this typical PWR of the GRS is similar to ANGRA 2 NPP, they are not identical, so that an adaptation of the modeling was necessary in order to apply it for analysis of severe accidents in NPP, ANGRA 2. For this reason, a considerable part of this study was dedicated for the adaptation of the modeling.

2.1 Description of the simulated accident scenario

In this study is presented a loss of coolant accident, in other words, Small Break LOCA (SBLOCA), in the hot leg. In this case, it is postulated that the rupture occurs in loop 1. The area of the rupture in the hot leg is 380 cm^2. The total flow area of the piping connected to the pressure vessel is 4418 cm^2, so that the area of the rupture is less than 10% of the piping area; reason why the rupture of 380 cm^2 is considered small.

The SBLOCA alone would not be sufficient to result in a severe accident, as long as the various plant safety systems were available. It was, therefore, necessary to add some aggravating conditions in order to cause the mentioned severe accident.

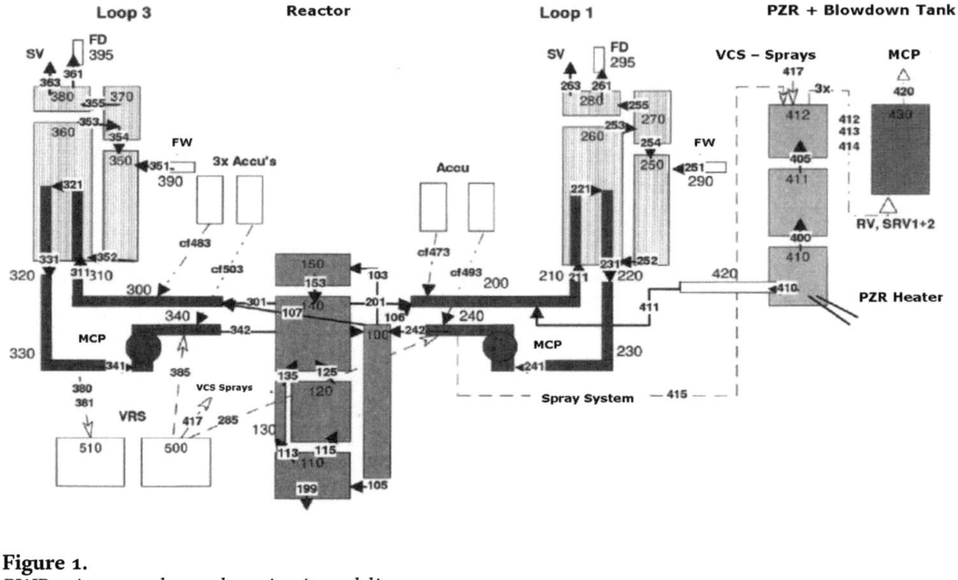

Figure 1.
PWR primary and secondary circuit modeling.

These conditions are conventionally known as boundary conditions. The postulated conditions are as follows:

a. Turbine bypass unavailable;

b. Condenser not available;

c. Loss of suction from sump and residual heat removal – RHR;

d. Injection of the Emergency Core Cooling System – ECCS from Refueling Water Storage Tank – RWST by Safety Injection Pump – SIP e do RHR available; and

e. All accumulators are available.

These boundary conditions, together with the primary circuit breakage, are sufficient to result in total core melting. Two mitigating measures were modeled in this study: Passive Autocatalytic Recombiner – PAR and Filtered Containment

Simulation time	Xe-133 m Bq	Cs-137 Bq	Ba-133 m Bq	Te-127 Bq
C1 – 2 h	2,11E+10	5,28E+01	1,40E+04	3,65E+06
C2 – 168 h	8,17E+19	1,56E+07	7,39E+10	2,00E+12

Table 1.
Source term for scenarios C1 and C2.

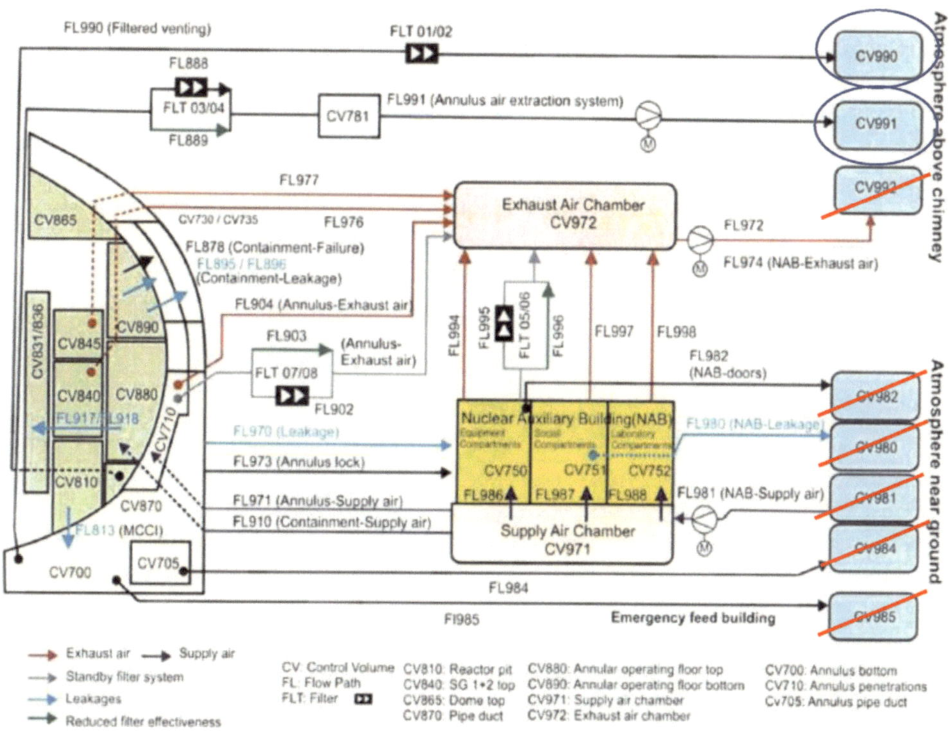

Figure 2.
Flow path (CV990 and CV991) of the radionuclides to the atmosphere.

Venting System – FCVS, with the purpose of evaluating their validity and efficiencies.

The inputs of the simulations performed for this work were elaborated using the input provided by GRS. This original input was modified according to the imposed conditions. It was assumed that the rupture occurs only in the Control Volume (CV) number 200, that is, in CV200 of loop 1, shown in **Figure 1**.

The flow path, given by FL070, has its opening and closing controlled by a CF660 control function. The setpoint for the valve opening is 7.0 bar and occurs after 168 h of simulation, but the setpoint for closing has not been implemented. It should be noted that the opening of the CV990 flow path, see **Figure 2**, occurs only after 168 h of simulation.

The flow path CV990 and CV991, see **Figure 2**, contains filters for aerosols and for fission products vapors so that some fraction of radionuclides are withdrawn through these filters when the vapors and aerosols are transported along of the flow path. A single filter can remove only one type of fission products, whether aerosols or vapors, but not both.

The radionuclide package of the MELCOR code contains a simple filter model which efficiency is defined by the global decontamination factor (DFG) determined by the user. The decontamination value assumed for aerosol filter was 1000, and for vapors of fission products, vapors were 100, which equated to 99.9 and 99.0% filtration, respectively.

2.2 Source term

The source term represents the radioactive inventory located in a system, equipment or component, which serves as a reference to evaluate the safety aspects in different conditions of operation of the reactor. It also represents one of the most important design bases for the study of installation performance, distribution of fission products in reactor systems, and in the environment in case of accidents.

Knowing the source term, there is the possibility of modeling radionuclide dispersion, calculating radiation concentrations and doses, as well as spatializing affected areas and environments. The source term of the present study is based on the radionuclides used in the MELCOR output, being this Xenon-133 m, Césiso-137, Barium-133 m, and Tellurium-127. **Table 1** shows the activities of the radionuclides for the scenarios C1 – simulation time of 2 h and C2 – simulation time of 168 h.

3. Characterization of the study area

The study area is placed in the south coast of Rio de Janeiro State, known region as "Costa Verde," in the city of Angra dos Reis, where the CNAAA is placed zone (23 k), latitude-UTM (7455581.00 m S), and longitude-UTM (555471.00 m E). The region geomorphology is extremely hilly, with quite steep slopes, high or negatives steepness and differences in elevations up to 800 m.

The geomorphology has two units of topography, one formed by ridge and scarps, and the other, by lowlands. The scarps has an average gap of 700 m and are dissected by half parallel valleys, which alternate with stretches of deep cutouts, between the rivers that flow down the mountain. **Figure 3** [14] shows the topography of the region.

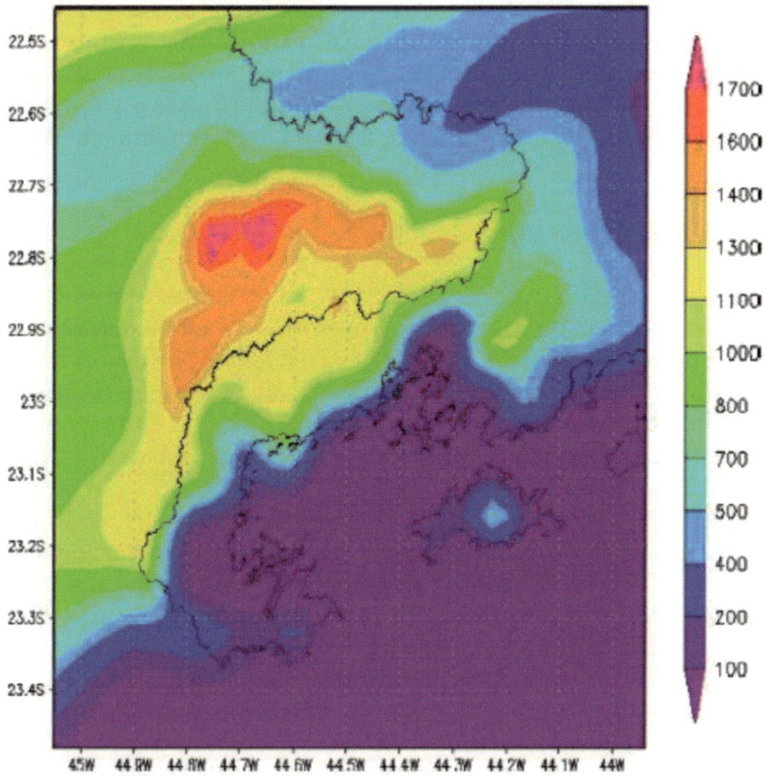

Figure 3.
Topography of the study area.

4. Methodology

The use of mathematical models facilitates the simulation process of transport mechanisms and pollutant deposition. These models provide a conservative theoretical estimate of the concentration levels of pollutants in the air, making it possible to evaluate the spatial and temporal evolution of these pollutants in the atmosphere.

4.1 Atmospheric model: WRF/CALMET

The WRF – Weather Research and Forecasting, is a numerical modeling system, developed for the weather forecasting and study of atmospheric phenomena of micro and mesoscale. Its development is the result of the collaboration between US research and government agencies centers: National Center for Atmospheric Research (NCAR), National Centers for Environmental Prediction (NCEP), National Oceanic and Atmospheric Administration (NOAA), US Departement of Defense, Oklahoma University, and Federal Aviation Administration (FAA).

The CALMET – California Meteorological Model, is a three-dimensional meteorological model that is integrated with the dispersion model – CALPUFF. The CALPOST is a post-processing package that makes possible to calculate the average concentration and deposition fluxes [15].

The CALMET is classified as a diagnostic meteorological model that incorporates meteorological observations and/or outputs of predictive meteorological models to produce, through objective analysis techniques, velocity, temperature, and other variables necessary for simulations with the CALPUFF model.

The CALMET requires that the meteorological and geophysical data are in specific formats before being used. The processing of these data is then performed with the aid of the preprocessors which prepare the data for assimilation in the CALMET processor.

4.2 Dispersion model: CALPUFF

The modeling of the atmospheric dispersion is a technique of simulation of the phenomena that occurs in nature, allowing to estimate the concentration of the pollutants in a set of points, based on a set of variables that influence them. The modeling of atmospheric dispersion is useful not only in the identification of emitting sources but also in the management of gaseous effluents and air quality, constituting one of the techniques of evaluation of air quality indicated by the environmental legislation.

The California Puff Model CALPUFF is a non-stationary puff model for dispersion simulations that can be used for a wide variety of applications in air quality modeling studies. The model was proposed and reviewed by Scire et al. [15] and has been adopted by the United States Environmental Protection Agency – EPA as a regulatory model for environmental impact studies covering distances of 50–300 km and including topography and complex meteorological systems.

The CALPUFF software is entirely public, designed to simulate release into the atmosphere, and is used to predict the effects of an accident, thus enabling effective emergency planning.

4.3 WRF/CALMET coupling

The choice of the WRF model configuration, as well as the domains, spatial resolution, and grid nesting, were made to obtain necessary meteorological data for the INPUT of the CALMET model. The initial and boundary conditions assimilated by the WRF are derived from the GFS (Global Forecasting System Model) of the National Centers for Environment Prediction (NCEP), whose spatial resolution is 0.5° (~55 km) and a time resolution of 3 h. To compose the GFS domain, both horizontally and vertically, a model for interpolation of the data is used. More details about this model can be found in KALNAY et al. [16]. The GFS data can be obtained for free from the electronic address. Link: https://www.ncdc.noaa.gov/data-access/model-data/model-datasets.

The meteorological data of the January month of 2009 was used for the simulation with the WRF. The January month data was chosen, due to being the most recent data obtained from Electronuclear for the four Angra towers.

The grid used by CALMET has a domain of 80 km and a cell number of 229 × 229. In the region of the NPP, the wind field data of the Electronuclear Towers was used, which radius of influence is 5 km.

4.4 Whole body dose calculation

The whole body dose is the contribution of the internal dose (inhalation or ingestion) added to the external dose (plume immersion) [17]. The calculations of the exposure pathways are in equations (Eqs. (1) and (2)), respectively. For the present study, the whole body dose analysis considered the plume concentration value and exposure time in the Emergency Planning Zones – EPZ, as shown in **Figure 4**.

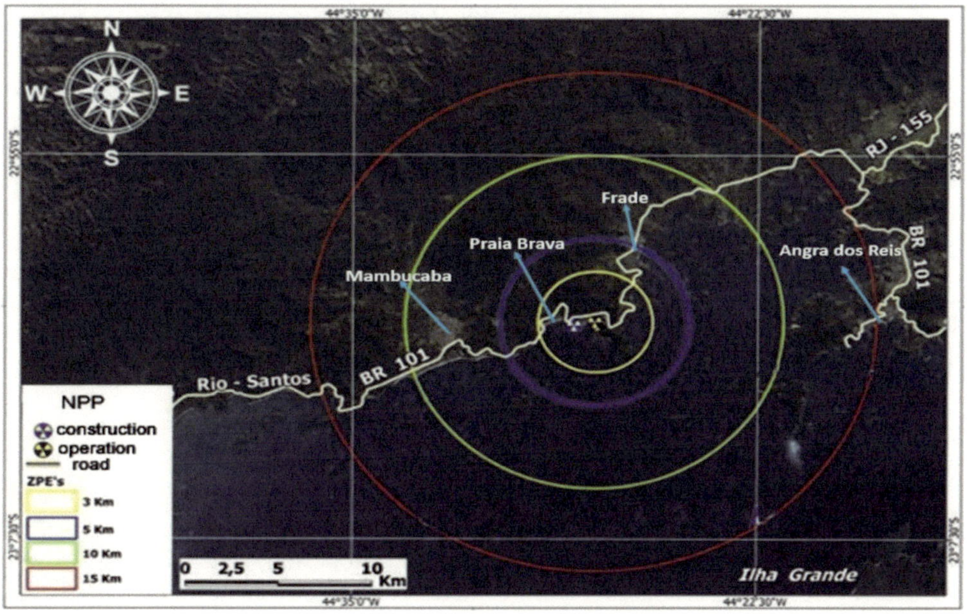

Figure 4.
Regions in EPZ that will be considered the dose calculation and measures protective.

4.4.1 Inhalation dose calculation

The inhalation dose calculation was based on the IAEA-TecDoc-1162 [18] document, as observed in Eq. (1).

$$D_{in} = \sum_i C_i * e(g)_i * B_r * T \quad (1)$$

Where:

$\boldsymbol{D_{in}}$ = Inhalation dose, Sv;
$\sum_i C_i$ = Concentration of each radionuclide – Bq/m^3;
$\boldsymbol{e(g)_i}$ = Inhalation dose coefficient – Sv/Bq (was used e(g)i for adult e(g) i > 17 years), according to Regulatory Position CNEN 3.01/011:2011 [19];
$\boldsymbol{B_r}$ = Breathing rate, whose value considered 1.5 m^3/h [18]; and
$\boldsymbol{T}$ = Exposure time – h (time the individual of the public will be exposed to the pollutant).

4.4.2 Immersion dose calculation

The immersion dose calculation was based on the IAEA-TecDoc-1162 [18] document, as observed in Eq. (2).

$$D_{im} = \sum_i C_i * e(g)_i * T \quad (2)$$

Where:

$\boldsymbol{D_{in}}$ = Immersion dose, Sv;
$\sum_i C_i$ = Concentration of each radionuclide – Bq/m^3;
$\boldsymbol{e(g)_i}$ = Immersion dose coefficient – Sv.m^3/Bq.h (was used e(g)i for adult e(g) i > 17 years), according to Regulatory Position CNEN 3.01/011:2011 [19]; and

T = Exposure time – h (time the individual of the public will be exposed to the pollutant).

5. Results

5.1 Pollutant transport model: CALPUFF

The simulations were performed on CALPUFF assuming a point source of emissions related to the chimney of the NPP Angra 2, which height considered was 155 m. The emission rate will follow scenarios C1 and C2, with the constant release rate of the radionuclides released into the atmosphere.

5.1.1 Radionuclides release for scenario C1

The simulation was performed from 05/01/2009 to 08/01/2009 at 06:00 h. It was considered that all the radionuclides of **Table 1**, after release, behave according to the region's wind field during the entire simulation period, 72 h. The respective plume concentration periods that were considered: 1, 3, and 72 h, for each EPZ, see **Tables 2–5. Figure 5** shows the wind field and radionuclides transport from **Table 1** for scenario C1.

5.1.2 Radionuclides release for scenario C2

The simulation was performed from 05/01/2009 to 08/01/2009 at 06:00 h. It was considered that all the radionuclides of **Table 1**, after release, behave according

Simulation time	Xe-133 m Bq/m^3	Cs-137 Bq/m^3	Ba-133 m Bq/m^3	Te-127 Bq/m^3
1 h	1,53E+04	3,76E-05	9,97E-03	2,60E+00
3 h	6,04E+03	1,48E-05	3,92E-03	1,02E+00
72 h	5,64E+02	1,21E-06	3,22E-04	8,39E-02

Table 2.
Radionuclides concentration in EPZ-3 km (Praia Brava).

Simulation time	Xe-133 m Bq/m^3	Cs-137 Bq/m^3	Ba-133 m Bq/m^3	Te-127 Bq/m^3
1 h	1,00E+04	2,37E-05	6,29E-03	1,64E+00
3 h	4,30E+03	9,22E-06	2,45E-03	6,37E-01
72 h	1,08E+03	2,15E-06	5,71E-04	1,49E-01

Table 3.
Radionuclides concentration in EPZ-5 km (Frade).

Simulation time	Xe-133 m Bq/m^3	Cs-137 Bq/m^3	Ba-133 m Bq/m^3	Te-127 Bq/m^3
1 h	1,54E+03	3,55E-06	9,41E-04	2,45E-01
3 h	9,34E+02	2,16E-06	5,74E-04	1,50E-01
72 h	1,60E+02	2,92E-07	7,75E-05	2,02E-02

Table 4.
Radionuclides concentration in EPZ-10 km (Mambucaba).

Simulation time	Xe-133 m Bq/m^3	Cs-137 Bq/m^3	Ba-133 m Bq/m^3	Te-127 Bq/m^3
1 h	1,09E+03	1,49E-06	3,96E-04	1,03E-01
3 h	8,29E+02	1,16E-06	3,07E-04	8,01E-02
72 h	2,21E+02	2,83E-07	7,51E-05	1,96E-02

Table 5.
Radionuclides concentration in EPZ-15 km (Angra dos Reis).

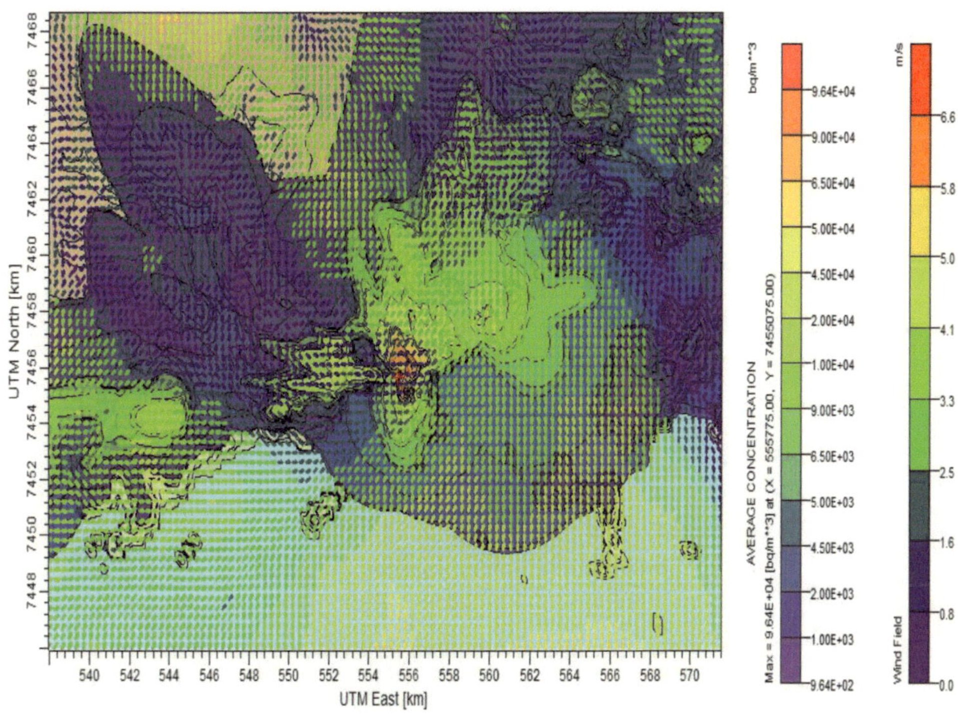

Figure 5.
Concentration and wind field for scenario C1.

Simulation time	Xe-133 m Bq/m^3	Cs-137 Bq/m^3	Ba-133 m Bq/m^3	Te-127 Bq/m^3
1 h	5,92E+13	1,11E+01	5,27E+04	1,43E+06
3 h	2,34E+13	4,37E+00	2,07E+04	5,60E+05
72 h	2,18E+12	3,58E-01	1,70E+03	4,60E+04

Table 6.
Radionuclides concentration in EPZ-3 km (Praia Brava).

to the region's wind field during the entire simulation period, 72 h. The respective plume concentration periods that were considered: 1, 3, and 72 h, for each EPZ, see **Tables 6–9**. **Figure 6** shows the wind field and radionuclides transport from **Table 1** for scenario C2.

5.2 Whole body dose in EPZ

- Whole body dose analysis for Scenario C1

Simulation time	Xe-133 m Bq/m^3	Cs-137 Bq/m^3	Ba-133 m Bq/m^3	Te-127 Bq/m^3
1 h	3,88E+13	7,01E+00	3,32E+04	8,99E+05
3 h	1,66E+13	2,72E+00	1,29E+04	3,49E+05
72 h	4,19E+12	6,36E-01	3,01E+03	8,16E+04

Table 7.
Radionuclides concentration in EPZ-5 km (Frade).

Simulation time	Xe-133 m Bq/m^3	Cs-137 Bq/m^3	Ba-133 m Bq/m^3	Te-127 Bq/m^3
1 h	5,97E+12	1,05E+00	4,97E+03	1,34E+05
3 h	3,62E+12	6,39E-01	3,03E+03	8,20E+04
72 h	6,18E+11	8,64E-02	4,09E+02	1,11E+04

Table 8.
Radionuclides concentration in EPZ-10 km (Mambucaba).

Simulation time	Xe-133 m Bq/m^3	Cs-137 Bq/m^3	Ba-133 m Bq/m^3	Te-127 Bq/m^3
1 h	4,23E+12	4,42E-01	2,09E+03	5,66E+04
3 h	3,21E+12	3,42E-01	1,62E+03	4,39E+04
72 h	8,57E+11	8,37E-02	3,96E+02	1,07E+04

Table 9.
Radionuclides concentration in EPZ-15 km (Angra dos Reis).

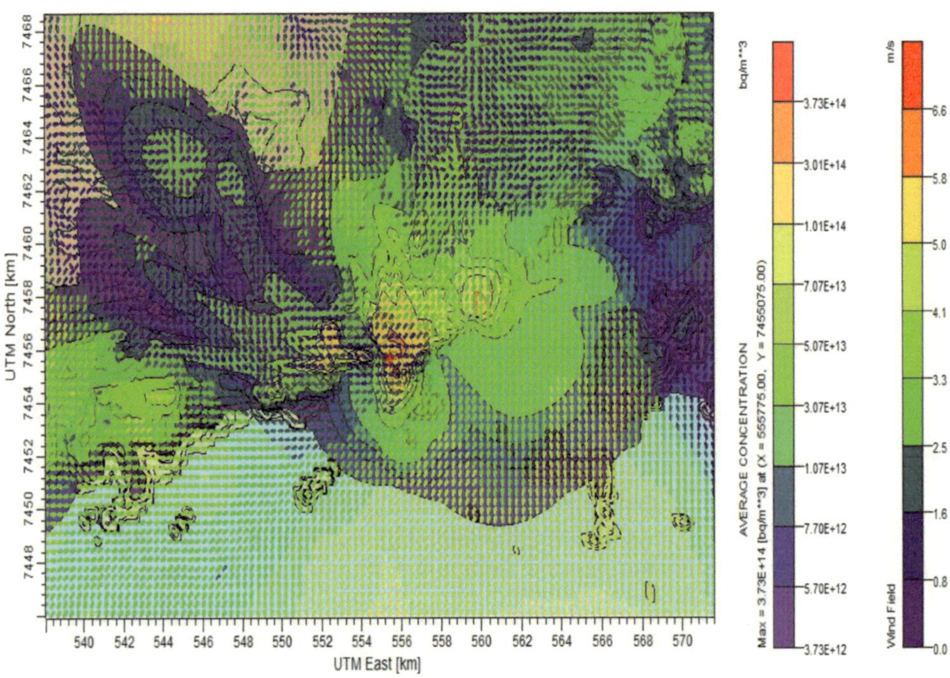

Figure 6.
Concentration and wind field for scenario C2.

C1 exposure time	EPZ 3 km Praia Brava (mSv)	EPZ 5 km Região do Frade (mSv)	EPZ 10 km Mambucaba (mSv)	EPZ 15 km Angra dos Reis (mSv)
1 h	7,06E-05	4,63E-05	7,12E-06	5,04E-06
3 h	8,38E-05	5,96E-05	1,29E-05	1,15E-05
72 h	1,94E-04	3,60E-04	5,31E-05	7,35E-05

Table 10.
Whole body dose in EPZ for scenario C1.

C2 exposure time	EPZ 3 km Praia Brava (mSv)	EPZ 5 km Região do Frade (mSv)	EPZ 10 km Mambucaba (mSv)	EPZ 15 km Angra dos Reis (mSv)
1 h	2,73E+05	1,79E+05	2,75E+04	1,95E+04
3 h	3,24E+05	2,30E+05	5,00E+04	4,44E+04
72 h	7,25E+05	1,39E+06	2,05E+05	2,84E+05

Table 11.
Whole body dose in EPZ for scenario C2.

For the dose calculation, were used the equations (Eqs. (1) and (2)), and the exposure time to individual members of the public was 1, 3, and 72 h. It was considered that these individual members of the public are not under any protection and the plume exposure is 100%. **Table 10** shows the dose values for each exposure time, as well as for each EPZ.

- Whole body dose analysis for Scenario C2

For the dose calculation, were used the equations (Eqs. (1) and (2)), and the exposure time to individual members of the public was 1, 3, and 72 h. It was considered that these individual members of the public are not under any protection and the plume exposure is 100%. **Table 11** shows the dose values for each exposure time, as well as for each EPZ.

6. Conclusions

According to the PEE/RJ [20] – External Emergency Plan of the State of Rio de Janeiro, in order to prioritize the risks and facilitate the planning and implementation of protection measures recommended by CNEN, the concept of EPZ was adopted. The Emergency Planning Zones were subdivided into circular crowns as shown in **Figure 4**.

According to the PEE/RJ, the preventive evacuation of the population constitutes an effective protection measure up to the distance of 5 km around the plant. From this distance, no additional benefit will be obtained with the preventive evacuation. Thus, to the EPZ 10 km and EPZ 15 km, it is preferable to recommend, in the short term, that the population remains sheltered. In this sense, the existing rays are classified as follows:

- Preventive action zones

EPZ 3 km: circumscribed area from 3 km centered on the nuclear unit of CNAAA, so the property area of ELETRONUCLEAR.

EPZ 5 km: circular crown, centered on the nuclear unit of CNAAA, with a 5 km outside radius and inner radius of 3 km. This EPZ 5 km is defined as the impact zone.

- Environmental control zones

EPZ 10 km: circular crown, centered on the nuclear unit of CNAAA with a 10 km outer radius and inner radius of 5 km.

EPZ 15 km: circular crown, centered on the nuclear unit of CNAAA with a 15 km outer radius and inner radius of 10 km.

According to a CNEN study [21], the protection measures for CNAAA can be divided as follows:

a. Area emergency: EPZ 3 km and EPZ 5 km (notification to the population to remain in residences or workplace, awaiting instructions) and EPZ 10 km and EPZ 15 km (notification to the population to keep on the alert for further instructions, keeping their normal activities); and

b. General Emergency: EPZ 3 km (population evacuation), EPZ 5 km (keep population sheltered), EPZ 10 km and EPZ 15 km (notification to the population to remain in the residences or workplace, awaiting instructions).

Based on the dose values for sheltered and evacuation of Regulatory Position CNEN 3.01/006: 2011 [22], whose values are 10 mSv for sheltered and 50 mSv for evacuation, will be recommended the protection measures for each EPZ for scenarios C1 and C2, as shown in **Tables 12** and **13**. The recommendations in **Table 12** were based on the projected doses of **Table 10** and the recommendations in **Table 13** were based on the projected doses of **Table 11**.

It is observed in **Table 12** that for the Preventive Action Zones (EPZ 3 km and EPZ 5 km) and Environmental Control Zones (EPZ 10 km and EPZ 15 km), the recommendations of the protective measures that best meet the dose levels is Sheltered in place. The Scenario C1 is characterized as Area Emergency, having as a measure of the local authorities the notification to the population to remain in their residences or workplace, awaiting future instructions.

C1 exposure time	EPZ 3 km Praia Brava	EPZ 5 km Frade	EPZ 10 km Mambucaba	EPZ 15 km Angra dos Reis
1 h	Sheltered	Sheltered	Sheltered	Sheltered
3 h	Sheltered	Sheltered	Sheltered	Sheltered
72 h	Sheltered	Sheltered	Sheltered	Sheltered

Table 12.
Recommendations for protection measures in EPZ for scenario C1.

C2 exposure time	EPZ 3 km Praia Brava	EPZ 5 km Frade	EPZ 10 km Mambucaba	EPZ 15 km Angra dos Reis
1 h	Evacuation	Evacuation	Evacuation	Evacuation
3 h	Evacuation	Evacuation	Evacuation	Evacuation
72 h	Evacuation	Evacuation	Evacuation	Evacuation

Table 13.
Recommendations for protection measures in EPZ for scenario C2.

It is observed in **Table 13** that for the Preventive Action Zones (EPZ 3 km and EPZ 5 km) and Environmental Control Zones (EPZ 10 km and EPZ 15 km), the recommendations of the protective measures that best meet the dose levels is Evacuation. The Scenario C2 is characterized as a General Emergency, having as a measure of the local authorities the evacuation of the population in EPZ.

In summary, it is possible to conclude that:

- Scenario C1 it is observed the occurrence of an Area Emergency, having as a protective measure in all EPZ, Sheltered in place.

- Scenario C2 it is observed the occurrence of a General Emergency, having as a protective measure in all EPZ, Evacuation;

- The impact zone that is currently 5 km, covering the Preventive Action Zones, for scenario C2, has the extension of this impact zone for distances beyond 5 km; and

- Analyzing the C1 and C2 scenarios, it is inferred that the faster the accident is mitigated, the lower will be the radiological consequences and therefore, the actions to the protective measures will be lighter.

Advance planning is essential to identify potential problems that may occur in an evacuation. The NRC case study cites the following aspects of planning as contributing to efficiency and effectiveness of evacuation [23]:

- High level of cooperation among agencies;

- Use of multiple forms of emergency communications;

- Community familiarity with alerting methods, the nature of the hazard and evacuation procedures;

- Community communication; and

- Well-trained emergency responders.

However, if environmental monitoring confirms that the population's exposure will extend beyond a few days, justifying other protection actions beyond shelter and evacuation, temporary or permanent resettlement should be considered [22].

Author details

André Silva de Aguiar*, Seung Min Lee and Gaianê Sabundjian
Nuclear and Energy Research Institute (IPEN/CNEN - SP), São Paulo, SP, Brazil

*Address all correspondence to: aguiargm@gmail.com

References

[1] Corey GR. A brief review of the accident at three mile island. IAEA Bulletin, Vol. 21, no.5. Available from: https://inis.iaea.org/search/search.aspx?orig_q=RN:11554960 [Accessed: 20 October 2019]

[2] Chernobyl's Legacy. Health, Environmental and Socio-Economic Impacts and Recommendations to the Governments of Belarus, the Russian Federation and Ukraine. Vienna: IAEA; 2003-2005

[3] Gauntt R, Kalinich D, Cardoni J, Phillips J, Goldmann A, Pickering S, et al. Fukushima Daiichi Accident Study. Sandia Report SAND2012–6173; 2012

[4] U.S.NRC – Nuclear Regulatory Commission. Severe Accidents. Available from: http://www.nrc.gov/reading-rm/doc-collections/nuregs/staff/sr1793/initial/chapter19.pdf [Accessed: 20 October 2019]

[5] Wooton RO, Cybuiskis P, Quayle SF. MARCH2 (meltdown accident response characteristics). In: Code Description and User's Manual. Washington, D.C.: Nuclear Regulatory Commission; 1984. (NUREG/CR-3988)

[6] Kim SH, Kim DH, Koh BR, Pessanha J, Siahmed EL-K, Podowski MZ, et al. The Development of APRIL.MOD2 – A Computer Code for Core Meltdown Accident Analysis of Boiling Water Nuclear Reactors. NUREG/CR-5157, ORNL/Sub/81–9089/3; 1988

[7] U.S.NRC – Nuclear Regulatory Commission. MELCOR Computer Code Manuals. Sandia National Laboratories. NUREG/CR-6119; 1990

[8] Gauntt RO, Cole RK, et al. MELCOR computer code manuals. NUREG/CR-6119, Rev. 1. USA: National Laboratory; 1998

[9] Gauntt RO, Cole RK, et al. MELCOR Computer Code Manuals. Vo1. 2: Reference Manuals, Version 1.8.5, Prepared by Sandia National Laboratories (SNL) for the U.S. Nuclear Regulatory Commission, Office of Nuclear Regulatory Research, NUREG/CR-6119, Vo1. 2, Re'v. 2; 2000

[10] Allison CM, et al. SCDAP/RELAPS/MOD3.1 code manual. Volume II: Damage Progression Model Theory, Technical Report NUREG/CR-6150, EGG-2720. USA: INEL; 1993

[11] Alliso NCM, et al. SCDAP/RELAPS/MOD3.2 Code Manual. Volume I-V, NUREG/CR-6150, INEL 96/0422, Revision 1; 1997

[12] MAAP4: Modular Accident Analysis Program for LWR Plants. Code Manual Vols. 1–4, Prepared by Fauske & Associates, Inc., Burr Ridge, IL, USA for the EPRI, Palo Alto, CA, USA; 1994

[13] Lamarsh JR. Introduction to Nuclear Reactor Theory. New york University: Addison-Wesley Publishing Company; 1975

[14] Aguiar AS, Lamego Simoes Filho FF, Alvim ACM, Pimentel LCG, Moraes NO. Station Blackout in unit 1 and analysis of the wind field in the region of Angra dos Reis. Annals of Nuclear Energy [Internet]. Elsevier BV; April 2015;**78**:93-103. DOI: 10.1016/j.anucene.2014.12.010

[15] Scire JS et al. A user's Guide for the CALPUFF Dispersion Model. (Version 5). Land O'Lakes, Florida, USA: Earth Tech. Inc; 2000

[16] Kalnay E, Kanamitsu M, Kistler R, Collins W, Deaven D, Gandin L, et al. The NCEP/NCAR 40-year reanalysis project. Bulletin of the American Meteorological Society. 1996;**77**(3): 437-470

[17] U.S. NRC – Nuclear Regulatory Commission. Dose Standards and Methods for Protection against Radiation and Contamination. USNRC Technical Training Center

[18] IAEA – International Atomic Energy Agency. Generic Procedures for Assessment and Response during a Radiological Emergency. TECDOC-1162

[19] CNEN – Comissão Nacional de Energia Nuclear. Coeficientes de Dose para Exposição do Público. Posição Regulatória; 2011

[20] SESDEC – Secretaria de Estado de Saúde e Defesa Civil. Plano de Emergência Externo do Estado do Rio de Janeiro. Para Caso de Emergência Nuclear nas Instalações da Central Nuclear Almirante Álvaro Alberto (CNAAA); 2008

[21] CNEN – Comissão Nacional de Energia Nuclear. Plano de Emergência Setorial – CNAAA. CGRN/DRS. Seminário Plano de Emergência – Marinha do Brasil; 2011

[22] CNEN – Comissão Nacional de Energia Nuclear. Medidas de Proteção e Critérios de Intervenção em Situações de Emergência. Posição Regulatória; 2011

[23] U.S. NRC – Nuclear Regulatory Commission. Identification and Analysis of Factors Affecting Emergency Evacuations. NUREG/CR-6864. Washington, D.C.; 2005

Chapter 7

Optimization of Cosmic Radiation Detection in Saline Environment

Valeriu Savu, Mădălin Ion Rusu and Dan Savastru

Abstract

Following the interaction of a neutrino with saline environment, the Cherenkov cone will be generated. The electromagnetic effect of the Cherenkov cone is perpendicular to the cone generator and it has the energy directly proportional to the neutrino energy. In the saline environment, neutrinos with very high energies (noise – 115 dBm) can be determined. Investigation of these neutrinos will lead to the construction of a Cherenkov detector. The construction of a Cherenkov detector involves the design and the construction of a very large number of detection elements and of cascade amplifiers. Another necessary condition is to know exactly the distribution of the dielectric parameters of the saline environment. In order to know the distribution of the dielectric parameters of the saline environment, it is necessary to make a map of their distribution. Under these conditions, the number of detection elements will be optimized and also the optimal position of the future Cherenkov detector will be determined. In this chapter, we will present the methodology of calculating the detection elements and a method to determine the dielectric parameters. Measurements of attenuation of the propagation of electromagnetic waves in this environment will be presented. We will detail how to optimize a Cherenkov detector.

Keywords: netrins, radiation, cone, Cherenkov, electromagnetic

1. Introduction

Following the interaction of a neutrino with saline environment the Cherenkov cone will be generated. This cone has the height in the prolongation of the neutrino's direction and the base of the Cherenkov cone is forming in the continuation of the neutrino's direction, keeping its angle at the top of the cone. The base of the Cherenkov cone moves further in the same direction as the neutrino that produced the Cherenkov Effect. The electromagnetic effect of the Cherenkov cone is perpendicular to the lateral surface of the cone and it has the energy directly proportional to the energy of the neutrino. It is this neutrino that produced this effect. By determining the energy and the direction of the neutrino that produced the electromagnetic effect of the Cherenkov cone, information about the phenomena in the Universe that generated this neutrino is discovered. In the saline environment, neutrinos with very high energies can be determined. These neutrinos provide information about the phenomena in the Universe that occurred at great distances from Earth. These distances are much larger than the distances at which the most efficient telescopes can work, so that the information obtained from neutrinos will increase the horizon of knowledge and contribute to the improvement of information about the Universe. Thus, we can say that this information makes a significant contribution in the field of astrophysics and astronomy.

The study of cosmic radiation began between 1911 and 1913. During this period, the Austrian Physicist Victor Hess, following balloon flights, measured the variation of ionization present in the air with the altitude [1]. The neutrinos carried by cosmic radiation have very high energies of the order ($10^{15} \div 10^{23}$) eV, those with energy between ($10^{15} \div 10^{21}$) eV can be detected in saline environment and those with energies greater than 10^{21} eV cross the terra.

The investigation of the interactions of high-energy neutrinos of cosmic origin in a dense environment (natural salt) will lead to the construction of a cosmic radiation observer in this environment. The phenomenon by which particles charged with high energies are detected due to the interaction with the environment is called the Askaryan effect and consists in the coherent emission of Cherenkov radiations in the radio frequency domain, through an excessive electrical charge that occurs during the development of an electron cascade in that environment. Cherenkov radiation occurs in the case of particles moving through an environment at a speed greater than the speed of light through that environment [2].

An avalanche of relativistic particles [3] represents the interaction between a very high-energy neutrino and a dense and dielectric environment (salt block). For neutrinos with energy greater than 10^{15} eV, only about 20% of it appears as a hadronic particles cascade, and this cascade has an electromagnetic component [3]. The electromagnetic cascade consists of electrically charged particles (about 70% of the particles) [4]. These particles contribute to the generation of the total electromagnetic energy of the cascade [4]. Particles with a speed of travel greater than the speed of light through a transparent and dense environment (the salt block) will produce the Cherenkov radiation effect (in our electromagnetic case) in this environment [2].

The phenomenon, by which the interaction with the environment can detect particles charged with high energies, is called the Askaryan effect and consists in the emission of Cherenkov radiation (in the case of particles moving through an environment at a speed greater than the speed of light through that environment) coherent in the radio frequency domain by the excess load that appears during the development of a cascade in that environment [2]. Determination of neutrinos with energies greater than 10^{12} eV can lead to the discovery of new astrophysical systems and new physical processes [3]. The direction, from which these very high energy neutrinos come from, is a direct indicator of the source that generates them, thus a cosmic radiation observer from a saline will have to fulfill this goal [3]. The result of the interaction of a very high energy neutrino with a dielectric, transparent and dense environment (salt block), is an avalanche of relativistic particles [3] which by Cherenkov effect will cause the information obtained to generate new aspects about astro-particles and they create the premises for a deeper understanding of the cosmic phenomena of high energy in the Universe [3]. These particles contribute to the generation of total electromagnetic energy in the form of a Cherenkov cone [4]. Knowing the effects related to the propagation of electromagnetic waves in dielectric environments with impurities (saline environment) [5], then, by eliminating the influences of the parameters of the propagation environment (saline environment), one can deduce the basic parameters of the flexible transmitting and receiving antennas. By performing a sufficient number of measurements, these basic parameters of the flexible transmitting and receiving antennas can be determined. Knowing these parameters, the detection of the electromagnetic radiation of the Cherenkov cone (which is known to be perpendicular to the generator of the cone) can be performed with much greater accuracy. Then it will be possible to determine the direction and energy level of the neutrinos generating the Cherenkov cone with the same precision level. Considering these aspects, the special importance of designing and realizing a complex system for determining the electrical parameters of the antennas for the detection of Cherenkov cone of electromagnetic radiation in saline environment, is deduced. The determination of the electrical parameters of the antennas for the detection of the Cherenkov cone of electromagnetic radiation in saline environment will be thought out so that it can determine these parameters in such environments. The basic parameters of the antennas [6, 7] will be determined: the radiation diagram, the directivity, the gain, the polarization, the input impedance, the frequency band, the effective surface, and the effective height.

In order to determine the Cherenkov cone in saline environment (the noise does not influence the energy measurement because the maximum noise level measured in saline environment is −115 dBm [8]), it is necessary to make a Cherenkov detector in this environment. The implementation of a Cherenkov detector in saline environment involves the design and construction of a very large number of detection elements together with the related devices and a very large number of cascade amplifiers as well [5]. Under these conditions the price of a Cherenkov detector in saline environment is very high. Another necessary condition (it is of particular importance since it can reduce the costs of producing a Cherenkov detector in saline environment) is the accurate knowledge and distribution of the dielectric parameters of the saline environment in the salt volume in which a Cherenkov detector will be made. The realization of a map of the distribution of the dielectric parameters of the saline environment in the entire volume of a salt rock implies the elaboration of a complex system for determining the dielectric parameters of the saline environment for the detection of the Cherenkov cone of cosmic radiation in this environment. With this system, measurements can be made in saline (on-site) environment in order to make this map. The use of this system in the measurements will increase the possibility to implement a Cherenkov detector in saline environment. Until now, this system has not been used in saline environment for the detection of cosmic radiation, which brings a novelty in the field. The novelty in the field of cosmic radiation detection in saline environment has led to patent applications A/00959/05.12.2016 [9], A/00404/07.06.2018 [10] and A/00354/12.06.2019 [11].

So far, a number of studies were carried out in different environments in order to perform a Cherenkov cosmic radiation detector. The studies were conducted in environments such as: air, ice, salt rock [12, 13], limestone rock, etc. For saline environment the SalSA detector is known, under water (ANTARES, Baikal, NEMO, NESTOR, AUTEC etc.), under ice (AMANDA, ICECUBE and RICE), for atmosphere (ASHRA, AUGER, EUSO and OWR), between soil and air (GLUE, Forte'NuTel and ANITA) [14]. The "Salt Sensor Array" (SalSA) detector has as reference parameters, 10 x 10 rows of square surfaces, placed 250 m horizontally between them for a depth of 2000 m and placed at 182 m vertically between them, with 12 knots per row and for each row 12 detection elements, resulting in 14,400 detection elements. In the SalSA (saline environment) project, a 250 m attenuation length of the electromagnetic waves was obtained for a frequency band of 100 MHz ÷ 300 MHz using antennas with horizontal and vertical polarity [15]. Cherenkov 3D type detector with a geometry $20 \times 20 \times 20$ where the number of sub-bands is 1 and the number of antennas with the same polarization type is 2 has a size of 500 m^3, uses 32,000 detection elements and a number of 400 wells in the salt block and it uses the neutron electron cosmic radiation detection system [5, 9]. In these studies, there was no question to determine the dielectric parameters of the environments, in which the measurements were performed.

In saline environment, two projects were carried that studied the way of detecting the Cherenkov cone of electromagnetic radiation in saline environment, but in these projects, there was no search for flexibility and adaptability of the antennas for the most accurate detection of the Cherenkov cone of electromagnetic radiation in saline medium.

The study of the detection of cosmic neutrinos began almost 20 years ago. Several specific telescopes have been developed that have attempted to identify these particles. The results were not the ones expected. On 23-02-1987, a radiation source of cosmic neutrinos was identified for the first time. This was called "Supernova 1987A" and opened a new stage in the theory of cosmos evolution. For ice detectors (ANITA) [16] with an SNR > 1 allowance, all events occurring in the frequency band (100 ÷ 1000) MHz can be considered detectable. In another paper dealing with the detection of cosmic radiation at the ice surface in Antarctica [17], it is mentioned that if SNR = 1 is considered, then the number of events can be estimated. Nor does this work address the reflections, attenuations, and characteristic of the antennas. Another paper dealing with the interaction of neutrinos (UHE) [18] and referring to a constant detector volume, does not take into account the effects related to the signal-to-noise ratio, antennas, propagation through the study environment, aspects that we want to achieve in this project. Due to an inhomogeneous distribution of impurities in the saline environment, a theoretical approach to the propagation phenomenon of electromagnetic waves in this environment cannot be realized [19, 20].

In order to obtain the most accurate dielectric parameters of the saline environment, it is necessary to improve the system of measurement and the determination of these parameters. In order to reach the proposed objective it is necessary to minimize the errors introduced by adapting the detection elements (transmission and reception antennas) to the saline environment (the electrical parameters of the antennas: the working impedances, the directional characteristics in horizontal and vertical plane, the gain, etc. of the transmitting and receiving antennas that are affected by the saline environment), it is necessary to make a band-pass filter with the lowest insertion attenuation resulting in a uniform bandwidth and it is also necessary to make an amplifier with the amplification as much as possible constant in the working band (central frequency 187.5 MHz, amplification band at 3 dB greater than the bandwidth filter by at least 10% and the amplification can compensate for the losses introduced by the connection cables).

The determination of the Cherenkov cone in saline environment presents as a result the determination of the energy, the direction and the sense that the neutrinos, which interact with a saline environment, possess. These neutrinos provide information about the phenomena in the Universe that occurred at great distances from Earth. These distances are much larger than the distances at which the most efficient telescopes can work, so that information obtained from neutrinos will increase the horizon of knowledge and will contribute to the improvement of information about the Universe. Thus, we can say that this information makes a significant contribution in the field of astrophysics and astronomy.

2. Data measurement systems in saline environment

The generation of radiation pulses that arise from the interaction between high energy neutrinos (Ultra High Energy, UHE) and a dense dielectric medium has been studied first by Askaryan [21], who also presented the first results based on laboratory tests.

Askaryan also identified several natural materials that can be used as neutrinos detectors: the salt blocks present in saline mines, the ice from polar region, and the soil of moon [22, 23]. It was proven that a solid block of salt is a very good candidate for such detectors, since it suffers important changes of its electrical properties, based on which, the neutrinos that pass through the block can be detected.

Based on the Askaryan effect [24–26] the radiation that passes through a dense dielectric generates a cone of coherent radiation in the radio or microwave frequency domain, known as Cherenkov radiation [27–29]. In order to detect this radiation, one has to determine the frequency domain in which those radio impulses have maximum intensity and the parameters of an antenna that can be used in a conventional receiver.

In an experimental setup with a particular configuration of transmitter and receiver antennas, one can measure the level and the range of the radiation generated and, based on those results, can evaluate the neutrinos energy. The system proposed in this paper consists in an Anritsu MS2690A signal analyzer, with an incorporated signal generator, coupled to the transmitting and receiving antennas [30].

With this system, the dielectric parameters of the saline environment are determined first and, by knowing these parameters, the distance of attenuation of the propagation of electromagnetic waves through the saline environment can be determined (the distance at which the module of the electromagnetic field decreases to 1/e). Thus, it is possible to determine, following a package of measurements for the vertical plane [8], and for the horizontal plane [31], the distribution of the attenuation of the electromagnetic waves through the saline environment (the map of the distribution of the electromagnetic waves in the saline environment leading to the determination of the optimal position of placement in a saline environment of a Cherenkov detector), the determination of the minimum number of detection elements, and the optimal position of their placement in saline environment [10]. Based on the use of dedicated software, one can determine the extreme situations of the generation of the Cherenkov cone outside the volume of the Cherenkov detector [11].

In order to determine the dielectric parameters of the saline environment, two methods were studied, a direct and an indirect one.

The direct method involves the injection of a radio frequency signal into the measuring medium (saline medium) in order to determine the electrical parameters of the radio frequency antennas. Thus, this method involves performing a measuring assembly. This will include a radio frequency signal generator nozzle that will inject the signal into an emission antenna, the electrical parameters of which are known, an antenna for the reception of the injected signal, the electrical parameters of which will be determined, a signal analyzer block received from the measuring antenna. The antennas will be introduced in saline environment.

According to IEEE standard no. 145–1983 [32], which states that "the antenna is a means of transmitting or receiving radio waves", i.e. the antenna is that part of a radio equipment that, by means of electromagnetic exchange of power with the environment, ensures communication between at least two telecommunication equipments. The antenna can also be regarded as an element that adapts between the environment and the receiver or transmitter. It actually performs a transformation of the power of the electromagnetic field into a signal received as electrical power. Also, the antenna transforms the electric emission power into the power of the radiant electromagnetic field [6, 33].

The transmitting and receiving antennas, from a constructive point of view, are identical. The basic parameters of the antennas are [6, 7]:

- radiation diagram,
- directivity,

- gain,
- polarization,
- the input impedance,
- the frequency band,
- the actual area,
- the effective height.

The radiation diagram of the antenna represents the space surface for which the vectors leaving the antenna towards this surface have the module proportional to the intensity of the radiation in the respective direction. The direction, in which the field intensity is zero, is called null. The region between two nulls is called the lob. The maximum of the lobe is called the level of the lobe and the direction, in which it is maximum, is called the orientation of the lobe. If we represent the lobes in relative sizes, then an antenna can have one or more level 1 (0 dB) lobes - called main lobes, – and less than 1 level lobes (negative in dB) – named secondary (lateral or auxiliary) lobes. The back lobe of an antenna (180° to the main lobe) is related to the main lobe (in dB) and it is called the front/back ratio of the antenna [6, 7].

We can define the radiation diagram of an antenna if we take into account the electric component module E for the electromagnetic field radiated by the antenna. The other parameters and their definitions are kept. A decrease in 3 dB of the electric field module represents a decrease of its times ($1/\sqrt{2} \approx 0.707$) [6, 7].

The maximum radiation intensity is given by:

$$P_{\Omega max} = \left|P_\Omega\left(\theta = \frac{\pi}{2}\right)\right| = \left|\frac{\eta_0 k_0^2}{32\pi^2}\left(\sin\frac{\pi}{2}\right)^2\right| = \frac{\eta_0 k_0^2}{32\pi^2} \tag{1}$$

and the relative radiation intensity is given by:

$$P_{\Omega \text{rel}} = \frac{|P_\Omega|}{P_{\Omega max}} = \frac{\frac{\eta_0 k_0^2}{32\pi^2}(\sin\theta)^2}{\frac{\eta_0 k_0^2}{32\pi^2}} = (\sin\theta)^2 \tag{2}$$

where: P_Ω represents the radiation intensity,
θ represents the angle under which the radiation intensity is determined,
η_0 represents the vacuum impedance,
k_0 represents the vacuum propagation constant.
From these equations, the radiation pattern of the antenna can be determined.

Directivity is the ratio of radiation intensity in a given direction to the average radiation intensity that is calculated for all directions in space. The average radiation intensity is calculated as the total radiated power divided by 4π. The approximate formula for calculating directivity is as follows [6, 7]:

$$D(\theta, \phi) = 10\, log\left[4\pi\frac{P_\Omega(\theta, \phi)}{P_{rad}}\right][dB] \tag{3}$$

The absolute gain of an antenna, for a given direction, represents the radiation intensity in that direction relative to the radiation intensity that could be obtained if the antenna would radiate isotropically all input power. The radiation intensity, corresponding to the isotropically radiated power, is equal to the ratio of the input power to 4π. For an approximate calculation, the formula [6, 7] can be used:

$$G_{max} \approx \frac{3 \bullet 10^5}{\theta_{\mathbb{E}}\theta_{\mathbb{H}}} \tag{4}$$

where: and represents the angular openings (in degrees) at 3 dB in the planes of vectors E and H.

The polarization of an antenna is determined by the polarization of the electromagnetic field radiated by it. The propagation of the electromagnetic field is given by a transverse plane wave (components E and H are perpendicular to each other and in turn are perpendicular to the propagation direction). The polarization of an electromagnetic field is determined by the curve of the vector E described in time at the observation point. This curve can be an ellipse (elliptic polarization), a circle (circular polarization) or a straight line (linear polarization). Apart from linear polarization, the other polarizations are characterized by the direction of travel of the curve (right or left) [6, 7].

The input impedance of an antenna is, in fact, the impedance presented at the antenna terminals. The impedance of the antenna is given by the ratio between the voltage and the current at the terminals or the ratio between the electrical and magnetic components determined at a conveniently chosen point. The formula for calculating the impedance of the dipole antenna of length l is much smaller than λ and traveled by a constant current I is as follows [6, 7]:

$$R_{rad} = P_{rad} \Big/ \left(\frac{1}{2}I^2\right) = \frac{\eta_0 k_0^2 l^2}{6\pi} \tag{5}$$

The frequency band is defined as "the frequency range, in which the antenna performance associated with a predetermined parameter, is maintained in a specified range" [6, 7].

The actual surface area of an antenna, for a given direction, is represented by the ratio of the power available at the antenna terminals, being considered as the receiving antenna and the power density for the plane wave incident in that direction. The electromagnetic wave and the antenna are considered to be adapted from each other in terms of polarization. If no specific direction is indicated, then the maximum antenna radiation direction is taken by default [6, 7].

The effective height of an antenna, with a linear polarization and receiving a plane wave from a given direction, represents the ratio between the voltage determined with the open circuit at the antenna terminals and the intensity of the electric field determined by the antenna polarization direction.

An issue that interests us is the input resistance of the dipole antenna. This antenna will be calculated to work in a saline environment. In order to determine the parameters of the antenna in saline environment, we must know the input resistance of the dipole antenna in the free space. In order to determine the input resistance of the dipole antenna, we will start from the cylindrical dipole, which is a direct materialization of the concept of thin wire antenna. The parameters of the cylindrical dipole are slightly different from those provided by a theoretical analysis. This fact is given by the condition imposed on the length of the dipole, which must be much larger than the diameter. But this condition is not strictly fulfilled.

Considering that the dipole radiates in the free space, we will have approximate formulas for calculating the input resistance. If we make the notation G = nπ, then [6, 7]:

$$R_{in} = \begin{cases} 20G^2 & 0 < n < 1/4 \\ 24.7G^{2,5} & 1/4 \leq n < 1/2 \\ 11.4G^{4,17} & 1/2 \leq n < 0.6366 \end{cases} \tag{6}$$

The behavior of the dipole antenna in dielectric mediums for propagating the electromagnetic waves is similar to the behavior in vacuum or air, except that the impedance and the calculation of the antenna arm lengths change according to the relative permittivity of the environment, in which the antenna is located. If a group of antennas is inserted into a salt block, then the input resistance and the length of the dipole antenna in λ/2 will be changed with the real value of the permittivity of the salt (ε_r = 5.981 + j0.0835) and the penetration depth of the waves. Electromagnetic will depend on tgδ, which is precisely the ratio between the imaginary and the real part of the permittivity.

The antenna parameters are influenced when the antenna passes from work in vacuum or air to work in environment with different permittivity of vacuum. Therefore, in calculating the antennas working in environments with different permittivity than the vacuum (generally higher), the permissibility of the environment, in which the antenna works is taken into account. Averages such as salt constitute an unconventional environment for antennas and therefore the dielectric parameters of the salt, for a frequency range between 100 MHz and 5 GHz, must be known.

The direct method is performed by a system of generation and analysis of radio signals in saline environment and it is made from an emission antenna, a receiving antenna, and a signal analyzer. Two pairs of antennas are used for two working frequencies f1 and f2, in order to determine the dielectric parameters of the salt, in order to determine the transfer of electromagnetic waves through the salt block, and in order to determine the electrical parameters of the radiofrequency antennas in saline environment.

The system is made of two identical antennas of the dipole type in λ/2 for the wavelength corresponding to the frequency *f*1 = 450 MHz and *f*2 = 750 MHz, each provided with a symmetrical and an impedance adapter of a transformer type with transmission lines mounted in the immediate vicinity of antenna, coaxial cables with small losses of type RG58LL, an impedance adapter of type CD, connecting cables of type CDF400 with small losses between signal analyzer Anritsu MS2690A (generator part with emission antenna), two antennas (one for transmission and one for reception), and the signal analyzer Anritsu MS2690A (analyzer part with the receiving antenna). The schematic diagram of the system, for generating and analyzing radio signals in saline environment, is shown in **Figure 1** [8].

The indirect method involves the determination of the electrical parameters of the radio frequency antennas in saline environment, knowing the electrical parameters of these radio frequency antennas in air and measuring their parameters introduced in a saline environment when the dielectric parameters of the saline environment are known. This method involves performing a measurement installation of the electrical parameters of the radio frequency antennas in the air and repeating the measurements in a saline environment with the same installation. For this, dielectric parameters of the saline environment must be known.

Knowing the conclusions of the measurements in a saline environment, we can determine the electrical parameters of the radiofrequency antennas in such an environment. Numerous measurements have been made in massive salt blocks by RADAR penetration technology (GPR) [34]. From the conclusions of the measurements, we mention:

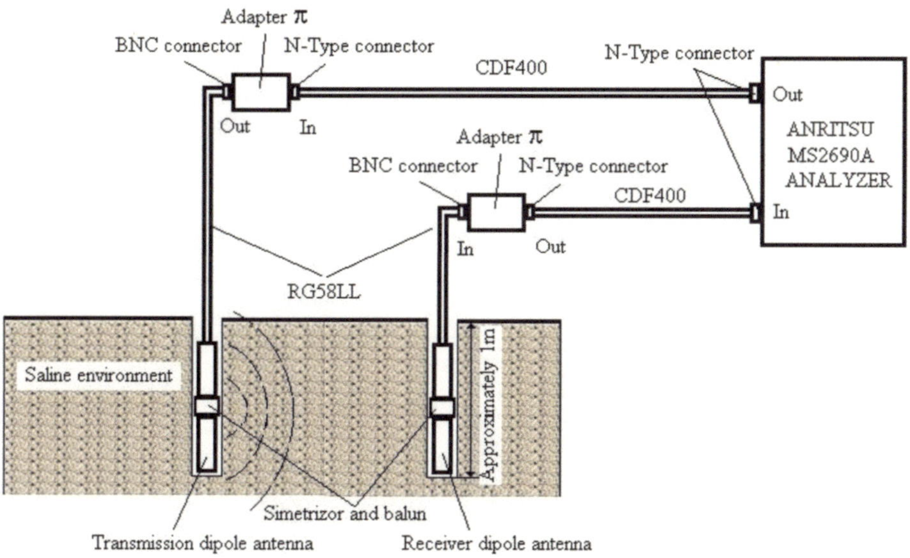

Figure 1.
The system of generation and analysis of radio signals in saline environment.

- the propagation of radio waves through a saline environment is not affected by scattering phenomena;
- no depolarization phenomena were observed;
- no significant dispersion phenomena were observed for the frequency range (0.1 ÷ 1) GHz.

Other empirical properties of salt blocks are [35]:

- a decrease in the tangent of the loss angle with frequency;
- the attenuation length is dependent on the percentage of impurities the salt block contains;
- no significant phenomena of double salt refraction were reported.

To perform the indirect method, the behavior of the antenna, introduced in a saline environment, will be analyzed when we have an interleaved element (air) between the antenna and the medium. By analyzing the following figure, we can determine the influence of the electrical parameters of the radio frequency antennas in saline environment when there is no perfect contact with this environment.

Figure 2 shows a cavity in a dielectric medium (salt), in which an emission or reception antenna (dipole antenna in λ/2) is introduced.

If we analyze the **Figure 1** where the cavity is cylindrical with the length L and the radius of the base of r = b and considering the continuity of the tangential component of the electric field ($E_0 = E_1$), we will find the equation below:

$$\frac{\Delta\omega}{\omega_0} = \frac{(\varepsilon_0(\varepsilon_r - 1))|E_{00}|^2 \pi b^2 L}{4W_0} \tag{7}$$

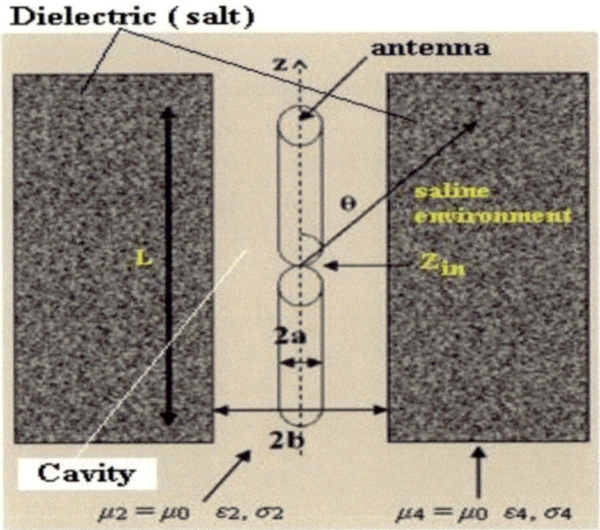

Figure 2.
Cavity in a dielectric medium (saline medium), in which a dipole antenna is inserted in λ/2.

where: $E_{00} = 1$ and $W_0 = \pi\varepsilon_0 L\int_a^b J_0(k_0 r)^2 r dr$ represents the energy found in the cavity. The resonant frequency of the empty cavity is:

$$\omega_0 = k_0 c = 2.405\left(\frac{c}{a}\right) \tag{8}$$

and $E_0 = J_0(k_0 r)$ is the energy in the empty cavity and J_0 represents the Bessel function of the first order for which $r = a$. Then, we will obtain:

$$\frac{\Delta\omega}{\omega_0} = 1.856(1 - \varepsilon_r)\left(\frac{b}{a}\right)^2 \tag{9}$$

If we use the last two formulas, then we can deduce the relative dielectric permittivity for the salt measurements:

$$\varepsilon_r = 5.981 + j0.0835 \tag{10}$$

The loss angle tangent is related to the attenuation coefficient of the field (α) and it represents the distance at which the electromagnetic field module decreases to (1/e). The formula, associated with the loss angle tangent, is [36]:

$$\tan\delta = \sqrt{\left[\frac{2}{\varepsilon'}\left(\frac{ac}{2\pi f}\right)^2 + 1\right]^2 - 1} \tag{11}$$

where: f represents the frequency and c is the speed of light in a vacuum. Then the attenuation length becomes:

$$L = \frac{1}{\alpha} = \frac{\lambda_0}{2\pi}\sqrt{\frac{2}{\varepsilon'\left(\sqrt{1 + \tan^2\delta} - 1\right)}} \tag{12}$$

or an approximate value:

$$L \cong \frac{\lambda_0}{2\pi(\varepsilon)^{\frac{1}{2}} \tan\delta} \tag{13}$$

where: λ_0 is the wavelength in vacuum.

For the study of the propagation of electromagnetic waves through saline environment it is necessary to know the dielectric parameters of the medium, through which they propagate (of the saline environment).

For this we will consider the propagation equation in linear, homogeneous, and isotropic environments for the electromagnetic waves [37–39]:

$$\nabla \bullet \bar{E} - \mu\sigma\frac{\partial \bar{E}}{\partial t} - \mu\varepsilon\frac{\partial^2 \bar{E}}{\partial t^2} = \nabla\left(\frac{\rho}{\varepsilon}\right) \tag{14}$$

and we consider the dissipative (absorbing) environment in which $\sigma \neq 0$.

We can assume that, if the environment contains free electric charge ($\rho \neq 0$), then it will exponentially decrease in time to zero.

We can determine the intensity of the wave at a certain depth z, which will be [37, 38]:

$$I(z) = \frac{1}{2}\sqrt{\frac{\varepsilon}{\mu}}\left(\bar{E}_0(z)\right)^2 = \frac{1}{2}\sqrt{\frac{\varepsilon}{\mu}}\left(\bar{E}_0(0)\right)^2 e^{-\beta z} \tag{15}$$

or otherwise:

$$I(z) = I_0 e^{-\beta z} \tag{16}$$

where: I_0 represents the intensity of the wave upon entering the environment (z = 0), β represents the absorption coefficient.

The salt from the mines of North America showed dielectric constants in the 5–7 range and the loss angle tangent between 0.015 and 0.030 at 300 MHz [40].

An important problem is to determine the penetration length of electromagnetic waves in saline environment (attenuation length.) Thus we will define the depth of penetration of the wave into the environment. We will note the distance d as representing this depth. This depth is the decrease of the intensity of the field e times from the initial one. Then the intensity of the wave at depth d becomes:

$$I(d) = I_0 e^{-\beta d} \tag{17}$$

where:

$$d = \frac{1}{\beta} \tag{18}$$

We will consider the case of an almost dielectric environment (σ – small, ε – big). In this case the ratio (σ/ε) will be much smaller than the unit [37, 38]:

$$\frac{\sigma}{\varepsilon\omega} \ll 1, \frac{\omega_c}{\omega} \ll 1 \text{ and } v \cong \frac{1}{\sqrt{\varepsilon\mu}} \tag{19}$$

which means that the electromagnetic wave has the same propagation speed for whatever its frequency is. This means that there is no dispersion (this is the case for salt).

Then the depth of penetration will be given by the relation:

$$d \cong \frac{1}{\sigma}\sqrt{\frac{\varepsilon}{\mu}} = \frac{1}{\sigma Z} \qquad (20)$$

where: Z is the impedance for the pulse of the wave.

In most practical cases $\tan\delta \ll 1$, so for the calculation of the penetration depth the formula is used (**Figure 3**):

$$d = \frac{3 \cdot 10^8}{2\pi f (\varepsilon_r)^{\frac{1}{2}} \tan\delta} \qquad (21)$$

The 36.79% percentage represents a 1/e decrease of the electromagnetic field in the dielectric environment (the incident field from which the reflected field is subtracted is taken into account).

The two methods do not differ much from each other. The difference is that, in the indirect method, there will be two packages of measurements. Starting from the package of measurements in air, continuing with the measurements in saline environment and knowing the dielectric and attenuation parameters of the electromagnetic waves of the saline environment, the electrical parameters of the radiofrequency antennas in saline environment can be determined by calculations. Taking into account these considerations, the indirect method can generate errors, because the determination of the electrical parameters of the radio frequency antennas in saline environment is based on the measurements of these parameters in the air (where small errors can occur), then measurements of these parameters are made in saline environment (where there also can occur small errors) and following the calculations, the errors can be added, which means a greater error. Thus, the indirect method involves high degree errors in determining the electrical parameters of radio frequency antennas in saline environment.

For the direct method, a system for measuring the electrical parameters of radio frequency antennas in saline environment will be used. The measurements being direct, we deduce that the errors are given only by these measurements (by the measurement system, analyzer – the generator part and the analyzer part). No additional calculations are required. So, the direct method is a method with smaller errors, although a measurement system, adapted to the saline environment is needed, compared to the indirect method that uses the same system of measurement in the air and in the saline environment.

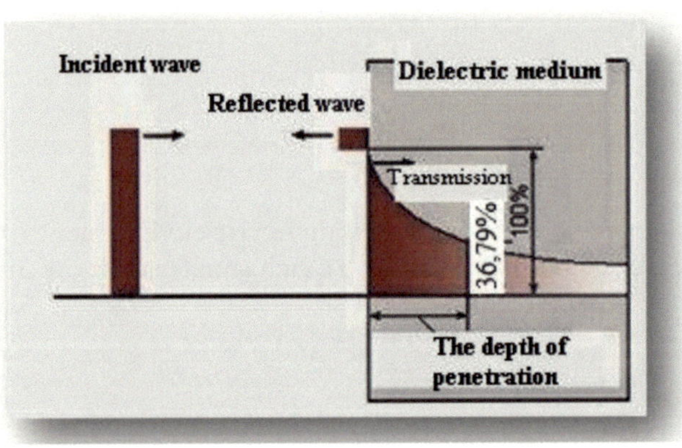

Figure 3.
Illustration of the penetration of electromagnetic waves in a dielectric environment [41].

3. Data collection in saline environment

In order to be able to collect the data from the saline environment, we will use the system presented in **Figure 1**. An important problem is the design of the antennas to work in the saline environment. The first problem, that arises, is the determination of the antenna length for working in saline environment.

Calculation of antennas [8]:

$$L_{a[\lambda/2]} = \frac{c}{2f\sqrt{\varepsilon_r}} \tag{22}$$

And it represents the antenna length in $\lambda/2$ [m] and $c = 3 \bullet 10^8$ m/s and f = the resonance frequency of the antenna [Hz], $\varepsilon_r = 5.981 + j0.0835$, taking into account the real part $\Re(\varepsilon_r) \cong 6$. **Figure 4** shows the shape and dimensions of the antenna determined by the above formula.

The antennas are made of Copper pipe with $\Phi = 6$ mm and have the following dimensions (**Table 1.**) for salt work.

A second problem that arises is the determination of the radiation resistance of the antennas in saline environment. As shown in **Figure 4**, the antenna is a dipole antenna in $\lambda/2$ and then the formula for calculating the radiation resistance is:

$$R_{rad} = 0.19397\sqrt{\frac{\mu_0}{\varepsilon_0 \times \varepsilon_r}} \tag{23}$$

where: $\mu_O = 4\pi 10^{-7}$ [N/m^2] and $\varepsilon_O = 8.859 \times 10^{-12}$ [F/m].

Then the radiation resistance is about 29.853 Ω. Following the analysis of **Table 1**, it is found that the antenna length is much smaller than λ and then we can say that the antennas are of Hertzian type and the radiation resistance of the antennas will be calculated with the formula:

$$R_{a\ sare} = \frac{2\pi}{3\sqrt{\varepsilon_{r\ sare}}} Z_0 \left(\frac{L_a}{\lambda}\right)^2 \tag{24}$$

Then, for salt work, a radiation resistance of about 13.5 Ω will be obtained. In these conditions, it is necessary to adapt the radiation resistance of the antennas to the characteristic impedance of the Anritsu MS2690A 50 Ω analyzer. For the

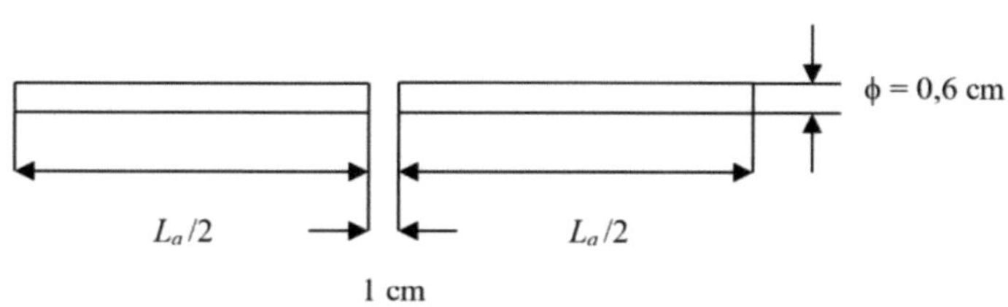

Figure 4.
Antenna dimensions.

f [MHz]	300	400	500	600	700	800	900	1000
L_a [m]	0.204	0.153	0.122	0.102	0.087	0.077	0.068	0.061

Table 1.
The dimensions of the transmitting and receiving antennas at different frequencies for working in saline environment.

frequencies $f1$ = 450 MHz and $f2$ = 750 MHz, $L_{a450MHz}$ = 0.136 m and $L_{a750MHz}$ = 0.082 m were obtained and for the radiation resistance the following values were obtained:

$$f_1 = 450MHz \Longrightarrow R_{a\ sare\ 450MHz} = \frac{1}{\sqrt{6}} \bullet 789.586 \bullet \left(\frac{0,136}{0,667}\right)^2 = 13.401\Omega \quad (25)$$

$$f_2 = 750MHz \Longrightarrow R_{a\ sare\ 750MHz} = \frac{1}{\sqrt{6}} \bullet 789.586 \bullet \left(\frac{0,082}{0,400}\right)^2 = 13.546\Omega \quad (26)$$

The frequency, at which the best propagation of electromagnetic waves, was determined in saline environment (Cantacuzino Mine from Slănic Prahova), is 187.5 MHz. The noise level in saline environment (Mina Cantacuzino from Slănic Prahova) is −115 dBm and at an impedance of 13.5 Ω, a noise level of 0.20662 μV is obtained. **Figure 5** shows the graph determined theoretically according to the distance of the variation of the radio frequency voltage level at the receiving antenna level. An electromagnetic emission event of a neutrino with the energy of 10^{18} eV that generated a Cherenkov cone in saline environment was taken into account in this graph.

Analyzing the graph in **Figure 5**, we determine that for a neutrino with the energy of 10^{18} eV that generated a Cherenkov cone in saline environment, a radio frequency signal at the terminals of a receiving antenna comparable to the noise level measured in saline environment will be produced as an effect at a distance of 50 m. So, for longer distances it is necessary that the energy of the neutrino be greater than 10^{18} eV (10^{23} eV).

Thus, a "Hardware system for detecting cosmic radiation of electron neutron type in salt" was designed [9]. This system is used to measure the level of attenuation of the electromagnetic waves introduced by the saline environment. Also with this system, the dielectric parameters of the saline environment can be determined in order to create a map with the distribution of the attenuations introduced by the saline environment. This system is, in fact, a radio detection station [5], SR_{mk} where m is equivalent to the Cartesian x coordinate, k is equivalent to the Cartesian y coordinate, and n is equivalent to the Cartesian z coordinate. Each radio station will include two antennas for horizontal polarization and two for vertical polarization, one impedance adjustment circuit for each antenna, one adder for each polarization, one band pass filter to select the desired spectral components, five amplifiers followed by one band pass filter (minus the last amplifier) with full chain amplification from the first band pass filter to the 120 dB anti-alloy filter, an anti-alloy filter, an analog-to-digital converter, a *FIFO* memory (first input – first output).

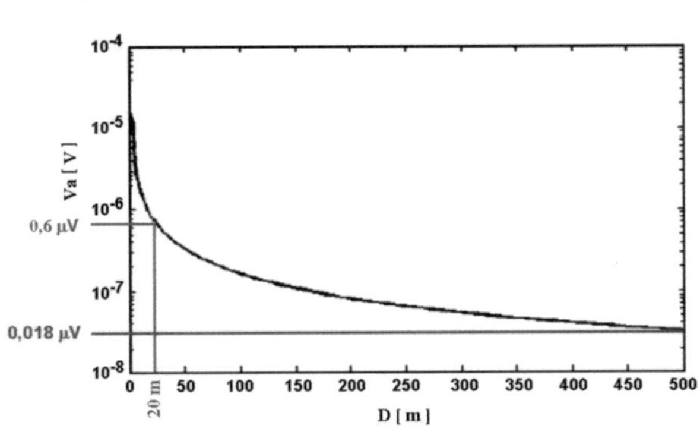

Figure 5.
The graph of the variation of the radio frequency voltage level at the receiving antenna level.

There follows a local processing station *SR* with a Wireless transceiver and, at the other end, another transceiver with a processing system to connect with the computer system. The total amplification of 120 dB is required to bring a signal of 0.6 μV (20 m) at the level of 0.6 V that can be processed by CAD_m (analog-to digital converter). It consists of the following blocks (**Figure 6**):

- A_{H1n} and A_{H2n} is the pair of antennas for horizontal polarization in the vertical group n;
- A_{V1n} and A_{V2n} is the pair of antennas for vertical polarization in group n on the vertical;
- CA_{H1n} and CA_{H2n} are the circuits for adapting the impedance of the antennas with the horizontal polarization to the impedance of the 50 Ω cable;
- CA_{V1n} and CA_{V2n} are the circuits for adapting the impedance of the antennas with the vertical polarization to the impedance of the 50 Ω cable;
- $\sum_{Hn}$ and $\sum_{Vn}$ are the sums of the signals coming from the antennas A_{H1n}, A_{H2n} respectively A_{V1n}, A_{V2n};
- FTB_{0m} is the first band pass filter to select the desired spectral components;
- A_{1m}, $FTB_{1m} - FTB_{4m}$, A_{5m} represents the 100 dB amplification chain together with FTB_{0m} and FAA_m;
- CAD_m is the converter from analog signal after amplification, to digital signal;
- $FIFO_m$ is the memory of the digital signal maintained as a buffer until the arrival of the trigger signal from the *P* + *Tx*/*Rx Wireless* (transceiver) system;
- *P* + *Tx*/*Rx Wireless* is the local information processing system that includes the hard and soft trigger circuit, the *Tx*/*Rx Wireless* transceiver and the receiving broadcast antenna, which is used to improve the noise signal ratio and to calibrate the system processing from *PC* computer.

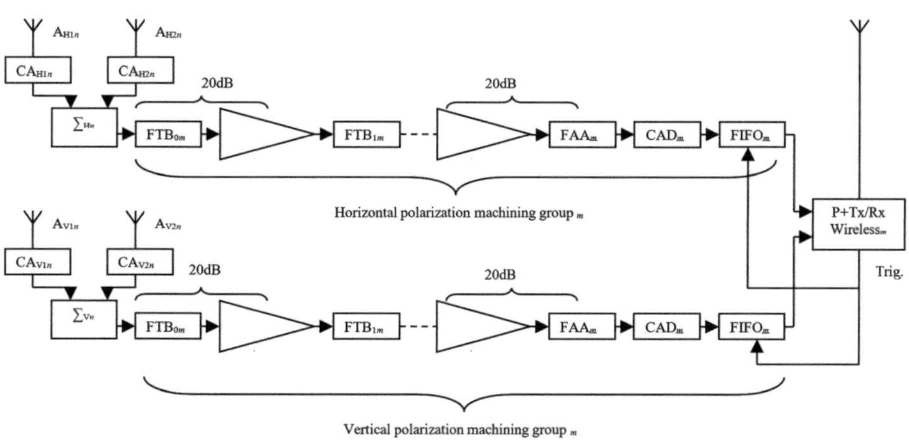

Figure 6.
Block diagram of the system for receiving, local processing and wireless transmission of the measured data to the computing system.

For the correct analysis of the data it is necessary that the temporal relation and the absolute value of the electric field be known, that is to say, all the instrumental errors must be corrected before working with the involved physical quantities. This implies a correction of the delays, which occur in the system, and a calibration of the amplitude.

For the correction of these events it is necessary that a well-known signal be present in all data and that it will provide us the necessary temporal information. There is no need for an absolute time scale, as the measurements are not compared to external events. For this reason, only the relative temporal delays between the antennas should be known.

It is necessary to determine the attenuations introduced by the connection cables.

We used two types of cables:

- type CFD400-E (blue) with a length of 5 m.
- type R-6763, O400 (black) with a length of 21 m.

The measured attenuations are presented in **Table 2**.

Following the measurements, the graph of variation of the power of a signal with a constant level measured at a fixed point at a distance of 20 m from the emission antenna was determined (**Figure 7**). The transmitting and receiving antennas were introduced in saline at a depth of 1 m from level 0.

Following the analysis of this graph, it is deduced that the attenuation of the electromagnetic waves is great for frequencies greater than 500 MHz, but it has a variation of about 20 dBm for a spectrum of 600 MHz. These attenuations fall for lengths of approximately 20 m. For the same signal levels introduced in the

f	Cable CFD400-E	Cable R-6763, O400
450 MHz	−0.31 dBm	−3.05 dBm
750 MHz	−0.6 dBm	−4.22 dBm

Table 2.
The attenuations measured on the connection cables used in the measurements in saline environment.

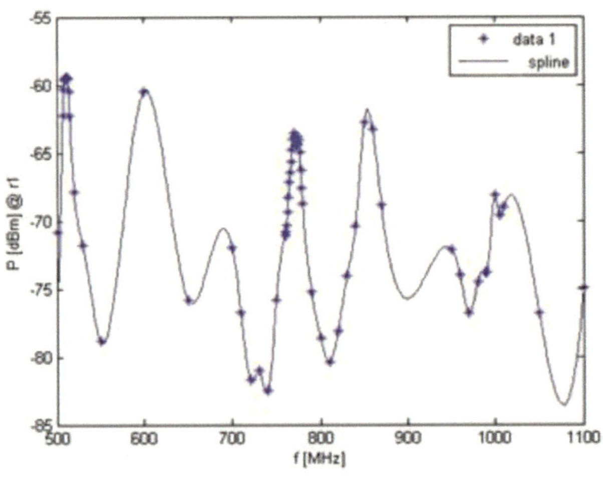

Figure 7.
The power received at a fixed point 20 m from the transmitting antenna for various frequencies emitted.

broadcast antenna placed at a depth of 1 m from level 0 and received with an identical antenna placed at the same depth, attenuations of about 50 m are obtained for the frequency of 187.5 MHz.

At the output of the system, the signal is processed by CAD_m, which is made with the RD-143 development board. It consists of an ADC083000 analog-to-digital converter (CAN), an amplifier for improving the signal/noise ratio achieved with the LMH6555 low-distortion differential amplifier, a tact signal generator, a Field Programmable Gate Array (FPGA) block, a local processor, a PLL loop frequency synthesizer, VCO oscillator made with LMX2531 LQ1500E, and a USB interface for direct communication with the PC made with CY7C6801BA.

The analog-numeric converter ADC083000 produced by National Semiconductor is an 8-bit converter that has a working power consumption of 1.9 W at a supply voltage of *Vcc* = 2.2 V, the maximum input signal on the Wine + and Wine− inputs is 2.5 V and the maximum conversion rate is 3GSPS (3 gigabytes per second). The resolution of the analog-numerical circuit, in this case, is:

$$Rez = 1LSB = \frac{V_{max}}{2^n} = \frac{2.5}{2^8} = 9.8mV \tag{27}$$

It also offers a bit error rate of the order of 10^{-18}.

Following the laboratory tests, it was found that the actual number of bits used by RD143 for quantization, around 187.5 MHz (the signal value from the CAN input – analog-to-digital converter) is 7.3 bits, which means that the noise introduced by the CAN is very small. In fact, this noise increases as the frequency of the signal processed by CAN increases.

Another important feature, determined by laboratory measurements, of this CAN is the low power consumption, reaching a consumption of 1.9 W at the maximum sampling frequency (3 GHz). Moreover, a linear characteristic of the power consumption, characteristic between 1.4 and 1.9 W. is observed. The signal processed by RD143 is on a frequency of 187.5 MHz.

The dipole antenna of the system was found to pick up the radio signal generated by the USRP (Universal Software Radio Peripheral). The signal reached in the signal processing unit (e.g. laptop) is a distorted sinusoidal signal with the fundamental on 187.49 MHz and the amplitude 66.23db (the input signal was −50 dB), which means an amplification of 116.23 dB (a close amplification of the theoretical one) (**Figure 8**).

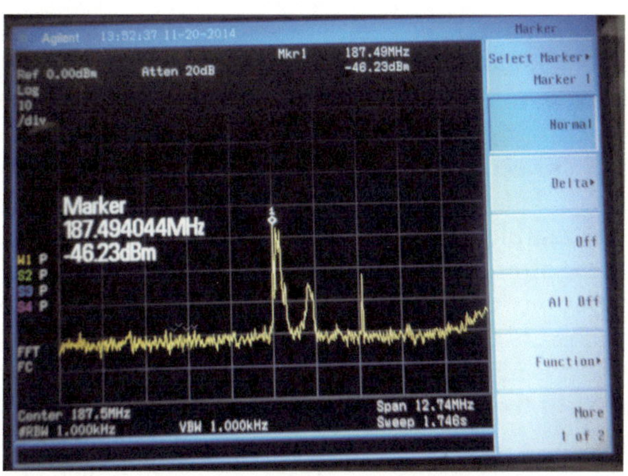

Figure 8.
The signal at the input of the RD143 digital processing system.

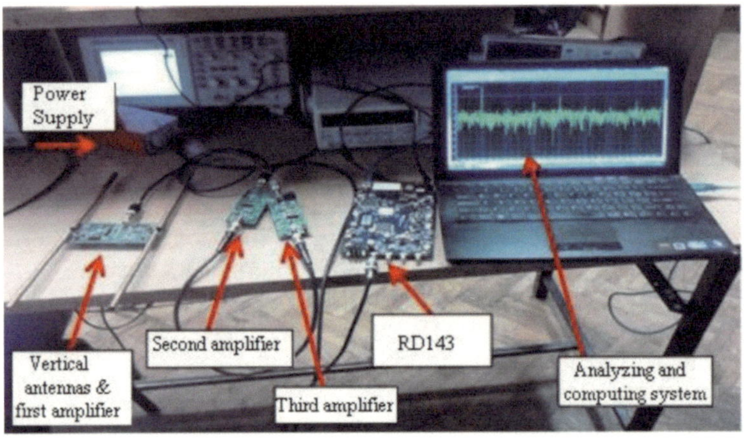

Figure 9.
*The system shown in **Figure 6** at the laboratory level.*

Distortions due to the existence of 3rd and 4th harmonics offer a distortion factor of 10%. The receiver design was performed for saline environment, medium with relative permeability different from that of the air. This could be one of the causes of a fairly large distortion factor. **Figure 9** shows the spectrum of the radio signal generated by the USRP and processed with the proposed experimental model.

4. Methods for determination and detection of Cherenkov cone in saline environment

Following the studies and articles published so far, it can be deduced that in order to make a Cherenkov detector in saline environment, many detecting elements and correspondingly many holes in saline environment, many chains of amplifiers (to bring the level detected by the workable TTL level detection (transistor-transistor logic), to compensate for losses on connection cables, etc.), many radio stations (SR_{mk}) etc. Under these conditions, an optimization of the number of detection elements of a Cherenkov detector in saline environment and implicitly of all corresponding component elements is required.

To optimize a Cherenkov detector, it is necessary to carry out a study in order to achieve the objective. Following the study it was concluded that the optimization of a Cherenkov detector in saline environment is necessary in order to determine the optimal positions by placing the detection elements for to obtain the maximum information. Thus, it is necessary to know the attenuation of electromagnetic waves in saline environment. This aspect involves a large number of measurements in the volume of the entire salt block in which the future Cherenkov detector will be placed. Because of the attenuation of the electromagnetic waves (the product of the interaction of a neutron with sufficiently high energy with a saline environment that generates the Cherenkov cone) is given by the dielectric permittivity of the saline environment, it is necessary to create a map with the distribution of the dielectric parameters of the saline environment. Knowing this map will determine the optimal positions of the detection elements and their number as well.

In these conditions, two patents have been proposed, which deal with the methods of determining the Cherenkov detector inside and outside the volume of the Cherenkov detector [10, 11].

4.1 Inside the volume of the Cherenkov detector

Determination of the Cherenkov cone inside the volume of the Cherenkov detector involves the design of a method to optimize the Cherenkov detector of electromagnetic radiation in the saline environment by determining the optimal points of placement of the detection elements and the Cherenkov detector in the saline environment, in order to minimize the number of measurement points and number of electromagnetic radiation sensing elements generated to reduce costs and simplify the measurement chain.

The first problem that occurs is the creation of a map with the distribution of the dielectric parameters of the saline environment. For this, a sufficiently large number of measurements of the dielectric parameters of the saline environment will be executed in order to interpolate and extrapolate the measurement results.

The problem solved by the optimization method of the Cherenkov detector of electromagnetic radiation in the saline environment removes the disadvantages of the Cherenkov detectors in the saline environment that have been proposed so far.

Thus the method minimizes the number of detection elements and implicitly of the measurement chain, being also an economical and much faster method, characterized in the fact that it determines the optimal points of placement of the detection elements for the determined volume of the Cherenkov detector in saline environment through iterations.

An iteration formula is used to obtain the optimal volume of the future Cherenkov detector placed in saline environment:

$$Lc_i^3 = p[i + \alpha(i - 1)] \tag{28}$$

where $\alpha, i \in \mathbf{N}$, $p \in \mathbf{R}$ and $p, \alpha, i > 0$; Lc_i^3 represents the length of the side of the cube with iteration i; i represents the number of the iteration; α represents a coefficient that is dependent on the attenuation length of the electromagnetic waves through saline environment and adjusts to whole values; p represents the iteration step and it is between 20 m ÷ 500 m.

This results in a Cherenkov detector consisting of at least two or more cube-shaped detectors in the cube and it also determines the optimal position of the future Cherenkov electromagnetic radiation detector in saline environment (**Figure 10**).

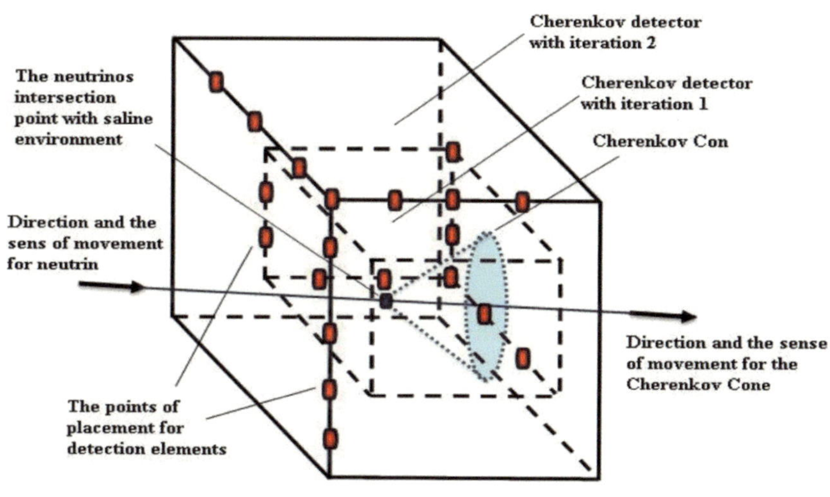

Figure 10.
An example of optimal placement of the detection elements of a Cherenkov detector for two iterations.

The method of optimization of the Cherenkov detector of electromagnetic radiation in the saline environment has, as first stage, the determination of the imprint of the saline environment, in which the future Cherenkov detector is placed, which is realized by measurements in order to determine the dielectric parameters of the environment and its attenuation length at the frequency of work set (187.5 MHz). In order to reach the first stage, measurements will be made to determine the propagation of the electromagnetic waves at different points of the volume of the environment in a vertical and horizontal plane [8, 31] and function of the measurement results, the measurements can be resumed or multiplied for to determine entirely the real distribution of the environmental attenuation for the propagation of the electromagnetic waves. These measurements will be performed using antennas whose electrical parameters (directivity characteristic, radiation resistance, loss resistance, antenna efficiency, front-to-rear ratio) are very well known to work in saline environment. These measurements will be performed horizontally and vertically, storing the data in a database, which from their processing they will lead to drawing a map with the distribution of the attenuation lengths of the electromagnetic waves. The second step consists in configuring the electrical parameters of the detection elements at this frequency (187.5 MHz), by determining the directivity characteristics, radiation resistance, loss resistance, efficiency and front-to-back ratio, for the horizontal and vertical plane. This data will be stored in another database, which represents the information regarding the detection elements of the Cherenkov detector. The determination of the electrical parameters of the detection elements of the Cherenkov detector will be carried out by resuming or multiplying the measurements so that the actual values of the electrical parameters for the detection elements can be determined as accurately as possible. The two databases (the environmental footprint and the electrical parameters of the detection elements) and the use of a dedicated software will determine the optimal placement points of the detection elements for the volume determined by each iteration. Thus, the minimum number of iterations can be determined to optimize the Cherenkov detector.

This method has the following advantages:

- Determination of the optimal positioning of the Cherenkov detector in the total volume of the saline environment;
- Determination of the optimal placement points of the detection elements for the volume determined by each iteration;
- Minimization of the number of detection elements and the measuring chain, which implies very low labor and material prices compared to the known methods and minimization of the number of wells necessary for the detector;
- Software processing time is short compared to other methods;
- Determination of Cherenkov Cone under real conditions.

This method determines the positions of the optimum measurement points, which lead to the minimization of the number of measurements, the number of electromagnetic radiation detection elements and implicitly of the measurement chain and it also determines the optimal position of the future Cherenkov detector in the total volume of the saline environment. All these aspects lead to cost reduction. In order to determine the Cherenkov cone in saline environment, the method requires the use of a dedicated software that uses a database containing the footprint of the saline environment for which a cosmic radiation detector is desired.

4.2 Outside the volume of the Cherenkov detector

The determination of the Cherenkov cone of electromagnetic radiation in the saline environment outside the volume of the Cherenkov detector is determined by placing at optimum points some detection elements outside its volume in all the x, y, z and positive and negative directions knowing the attenuation fingerprint in the electromagnetic wave field of the saline environment in order to determine the possible Cherenkov Cones that could form outside the detector volume depending on the energy determined by the detection elements in the vicinity of the detector.

So far, no method for the determination of a Cherenkov Cone outside the detector volume is known, even though some of the energy emitted by the cone reaches the detector elements of the detector.

This method determines the Cherenkov cone regardless of the position in which it is generated outside the detector volume, being an economical and predictable method, because outside the detector there are a minimum number of detection elements placed in optimal positions determined by their placement surfaces (planes) and obtained by using the formula:

$$Lc_{ex_i}^3 = p[1 + i(\alpha + 1)] \tag{29}$$

where: $\alpha, i \in N^*, p \in R$ şi $p, \alpha, i > 0$; $Lc_{ex_i}^3$ represents the length of the side of the cube delimited by the planes outside the volume of the Cherenkov detector; i represents the number of the iteration; α represents a coefficient that is dependent on the attenuation length of the electromagnetic waves through saline environment and it adjusts to whole values; p represents the iteration step and it is between $20 \div 500$ m and it is chosen larger than the iteration step of the Cherenkov cone determination resulted inside the Cherenkov detector volume (**Figure 11**).

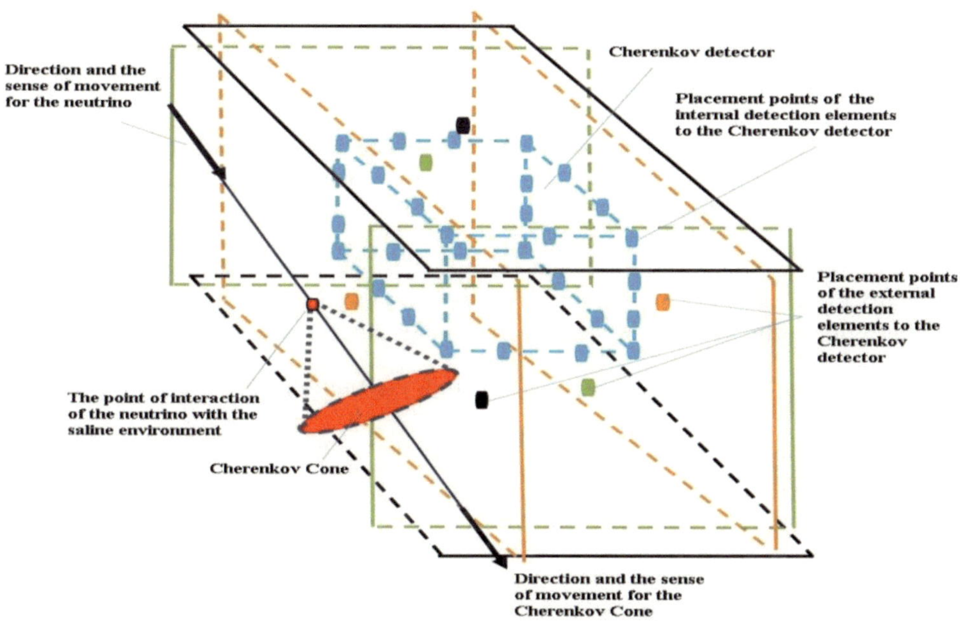

Figure 11.
An example of determination of the spatial positions of the placement plans of the detection elements external to the volume of the Cherenkov detector for a higher iteration of the determination of the volume of the Cherenkov detector in saline environment.

The method is based on the determination of the attenuation fingerprint of the saline environment (the map of the spatial distribution of the electromagnetic waves attenuation in the saline environment) in the field of the electromagnetic waves and it leads to the increased probability to determine the generation of the Cherenkov Cone from outside the detector volume by determining the energy levels measured by the external detection elements providing that they are higher than the energy levels measured by the detection elements inside the Cherenkov detector.

The determination of the Cherenkov cone in saline environment outside the volume of the Cherenkov detector consists in determining the optimal position of placement of the external detection elements, using a dedicated software.

For this, a minimum number of detection elements are placed outside the detector volume, which have very well established positions on the external surfaces of the detector, calculated by a higher iteration ratio than the detector volume calculation in all positive, negative x, y, z directions (Eq. (29)).

Then the attenuation of the saline environment (the map/fingerprint of the attenuation of the electromagnetic waves in the saline environment) will be calculated in the field of the electromagnetic waves [5, 8, 10, 31] as a result of measurements made outside the Cherenkov detector volume. In order to determine the Cherenkov cone outside the Cherenkov detector volume, the energy levels given by the sensing elements located outside the detector volume will be measured and if the energy measured by the external sensing elements is greater than the energy measured by the internal elements of the Cherenkov detector, then it is decided that a real Cherenkov cone outside the volume of the Cherenkov detector was generated. Thus we obtain the position in space of the Cherenkov Cone generated outside the detector in real situations using the dedicated software.

The method has the following advantages:

- determination of the Cherenkov cone generated outside the Cherenkov detector volume;

- determination of the optimal positioning of the detection elements outside the volume of the Cherenkov detector in the saline environment by using a dedicated software;

- it establishes the plans (surfaces) for placing the detection elements outside the Cherenkov detector volume obtained by an iteration higher than the detector volume determination;

- it minimizes the number of external detection elements and it optimizes the measurement chain, which implies very low labour and material prices and it minimizes the number of wells required outside the detector;

- it minimizes the data processing time with the help of the dedicated software;

- it determines the Cherenkov cone generated outside the detector under real conditions;

- this method can be used for any type of environment as long as the environmental mitigation footprint in the field in which the Cherenkov detector works in the respective environment does not change during the determination period.

The application of the method for the determination of the Cherenkov cone in saline environment outside the volume of the Cherenkov detector requires three steps prior to the method.

The first step consists in determining the dielectric parameters of the saline environment in which the detection elements from outside the volume of the Cherenkov detector are to be located and it represents precisely the footprint of the respective saline environment.

In the second stage, for the working frequency of the detection elements in the saline environment (187.5 MHz), the directivity characteristics, the radiation resistance, the loss resistance, the efficiency, and the front-to-back ratio are determined.

The third stage consists in processing the real data obtained when a Cherenkov cone is generated as a result of an interaction of cosmic radiation of the neutron nature with the saline environment in which the whole system (the Cherenkov detector and its external sensing elements) is located together with the two databases from the previous stages and depending on the energy levels measured by the external and internal elements the spatial position, in which the Cherenkov Cone was generated, is deduced. If the energy measured by the external sensing elements is greater than the energy measured by the internal elements then the Cherenkov Cone was generated outside the detector volume.

5. Conclusions

The information, "decoded" from the analysis of the electromagnetic energy generated by the Cherenkov cone (which is in a directly determined relation by the energy of the neutrino, which produced the Cherenkov phenomenon), are transmitted by the nuclear phenomena (fusion, fission, nuclear diffusion), which took place in the Universe at astronomic distanced (much larger than the detection possibilities known so far). This information brings an important contribution to the knowledge of the Universe.

The determination of the Cherenkov cone in salt spray (in salt spray the neutrinos with energies of the order $10^{12} \div 10^{23}$ eV are determined, which represent phenomena in the Universe that are generated by solar, galaxies, quasars, pulsars etc. systems), implies the implementation of a system, which measures the distribution of the density of the environment dielectric permittivity, which occurs in order to reduce the electromagnetic waves generated following the interaction between the environment with a neutrino. Thus a map, of the distribution of the "attenuation lengths" of the electromagnetic waves in the environment in which the measurement were done, is carried out.

The maximum distance between the detection elements placed in salt spray at Slănic Prahova is given by the noise level, which was measured here (−115 dBm) and it is of 50 m (0.2 μV, the graph from **Figure 5**). The detection of this level requires amplifications of about 129.5 dB on the frequency 187.5 Mhz (120 dB + 9.5 dB or an amplification of 3×10^6) in order to bring the signal at digital processing level with a DAC system (digital-analog converter). Thus we deduce that the attenuation length of the saline spray determines the placement points of the detection elements of a Cherenkov detector.

In order to minimize the costs of implementation of a Cherenkov detector for the determination and detection of the Cherenkov cone in saline spray (and in other environments where a map of the distribution of the spatial density of the dielectric permittivity in the volume of the entire environment, can be carried out), we need two stages: - the implementation of the map of the spatial density distribution of the dielectric permittivity in the volume of the entire salt spray and the determination of the optimum number of detection elements of the future Cherenkov detector and their optimum spatial placement position. In this regards, two methods for the determination and detection of the Cherenkov cone in salt spray are noticed: - inside and outside the volume of the detector.

The Cherenkov cone in salt spray is generated following the interaction of a UHE neutrino (Ultra High Energy, $10^{12} \div 10^{23}$ eV) with the saline environment. The detection of the information generated by the Cherenkov cone in salt spray implies knowing the energy, the direction, and the direction of travel of the neutrino, which interacted with this environment. The generation of UHE neutrinos may be due to some nuclear-related phenomena, which have a very high energy and give these neutrinos energies equivalent to the phenomena and provide information about these violent phenomena in the Universe. Thus, we can determine the nuclear phenomena in the Universe.

Acknowledgements

This work has been carried out on the Core Programme of the Romanian Ministry of Education and Research, National Authority for Scientific Research, PN-19-18 (18N/08. 02. 2019).

Notes/thanks/other declarations

We thank the Professors from the Faculty of Electronics, Telecommunications, and Information Technology at the Polytechnic University of Bucharest, Romania: Octavian Fratu, Alina Mihaela Bădescu, Alexandru Vulpe, and Răzvan Crăciunescu for the support and the materials made available.

Author details

Valeriu Savu, Mădălin Ion Rusu* and Dan Savastru
National Institute of Research and Development for Optoelectronics INOE 2000, Magurele, Romania

*Address all correspondence to: madalin@inoe.ro

References

[1] Hess VF. Convection phenomena in ionized gas-ion winds. Physikalishce Zeitschrift. 1912;**1912**:1084-1091

[2] Engel R, Sekel D, Stanev T. Neutrinos from propagation of ultra-high energy protons. Physics Review. 2001;**D64**: 093010

[3] Bădescu A. Radio technologies used in cosmic particle detection [Doctor Thesis]. Romania: The Polytechnic University of Bucharest; 2011

[4] Carroll B, Ostlie D. An introduction to modern astrophysics. In: Pearson, editor. San Francisco: Addison-Wesley; 2007. p. 18

[5] Savu V. Contributions to the selection and processing of radio pulses produced by cosmic radiations in the specific conditions of an observer built in a saline environment [Doctor Thesis]. Romania: The Polytechnic University of Bucharest; 2014

[6] Bogdan I. Antene şi propagare. Iaşi: Casa Venus; 2007

[7] Available from: http://www.et.upt.ro/admin/tmpfile/fileP1290448926file4ceab01e0c149.pdf

[8] Savu V, Marghescu I, Fratu O, Halunga S, Bădescu A. Antenna design for electromagnetic waves propagation studies through the salt ore. UPB Scientific Bulletin, Series C. 2013;**75**(2): 143-156

[9] Valeriu S, Octavian F, Răzvan-Eusebiu C, Viorica HS, Răzvan-Alexandru V, Carmen V. Hardware system for detecting cosmic radiation of electron neutron type in salt. Patent application A/00959/05.12.2016

[10] Valeriu S, Ion RM, Roxana S, Dan S. Method of optimization of the Cherenkov detector of electromagnetic radiation in saline environment. Patent application A/00404/07.06.2018

[11] Ion RM, Valeriu S, Dan S. Method of determining the Cherenkov cone in saline environment outside the volume of the Cherenkov detector. Patent application A/00354/12.06.2019

[12] Chiba M, Kamijo T, Yasuda O, Chikashige Y, Kon T, Takeoka Y, et al. Salt neutrino detector for ultrahigh-energy neutrinos. Physics of Atomic Nuclei. 2004;**67**(11):2050-2053

[13] Chiba M, Kamijo T, Kawaki M, Athar H, Inuzuka M, Ikeda M, et al. study of salt neutrino detector. AIP Conference Proceedings. 2001;**579**:204-221

[14] Connolly A. The radio Cerenkov technique for ultra-high energy neutrino detection. Nuclear Instruments and Methods in Physics Research Section A. 2008;**595**:260-263

[15] Available from: http://slideplayer.com/slide/7242558/

[16] Frichter G, Ralston J, McKay D. On radio detection of ultra high energy neutrinos in the Antarctic ice. Physical Review D. 1996;**53**:1684

[17] Provorov A, Zheleznykh I. Radiowave method of high energy neutrino detection: Calculation of the expected event rate. Astroparticle Physics. 1995;**4**(1)

[18] Gandhi R, Quigg C, Reno M, Sarcevic I. Neutrino interactions at ultrahigh energies. Physical Review D. 2006;**58**:093009

[19] Sjoberg D. Determination of propagation constants and material data from waveguide measurements. Progress In Electromagnetics Research B. 2009;**12**:163-182

[20] Hill DA. Fields of horizontal currents located above the earth. IEEE Transactions on Geoscience and Remote Sensing. 1989;**GE-26**(6)

[21] Saltzberg D, Gorham P, Walz D, et al. Observation of the Askaryan effect: Coherent microwave emission from change asymmetry in high energy particle cascades. Physical Review Letters. 2001;**86**:2802-2803

[22] Dagkesamanskii RD, Matveev VA, Zheleznykh IM. Prospects of radio detection of extremely high energy neutrinos bombarding the Moon. Nuclear Instruments and Methods in Physics Research A. 2010. In press

[23] Gorham PW et al. Experimental limit on the cosmic diffuse ultrahigh-energy neutrino flux. Physical Review Letters. 2004;**93**:041101

[24] Gorham PW et al. Accelerator measurements of the Askaryan effect in rock salt: A roadmap toward Teraton underground neutrino detectors. Physiological Reviews. 2005;**D72**: 023002

[25] Gorham PW et al. Observations of the Askaryan effect in ice. Physical Review Letters. 2007;**99**:171101

[26] Askaryan GA. Radiation of volume and surface compression waves during impingement of a nonrelativistic electron stream at the surface of a dense medium. Soviet physics - Technical physics. 1959;**4**(2):234-235

[27] Luo C, Ibanescu M, Johnson SG, Joannopoulos JD. Cerenkov radiation in photonic crystals. Science. 2003;**299**: 368-371

[28] Coleman SR, Glashow SL. Cosmic ray and neutrino tests of special relativity. Physics Letters B. 1997;**405**:249

[29] Zloshchastiev KG. Vacuum Cherenkov effect in logarithmic nonlinear quantum theory. Physics Letters A. 2010;**375**:2305-2308

[30] Badescu AM et al. Radio technique for investigating high energy cosmic neutrinos. Romanian Reports in Physics. 2012;**64**(1):281-293

[31] Savu V, Fratu O, Rusu MI, Savastru D, Tenciu D, Vulpe A, et al. Determination of the electromagnetic wave propagation for the detection of the Cherenkov radiation cone in salt environment. UPB Scientific Bulletin, Series A: Applied Mathematics and Physics. 2018;**80**(1):251-260

[32] IEEE Standard Definitions of Terms for Antennas, IEEE Std 145-1993 (Revision of IEEE Std 145-1983)

[33] Available from: http://www.scritube.com/stiinta/fizica/ECUATIILE-LUI-MAXWELL-PROPAGAR24262.php

[34] Unterberger R. Radar and sonar probing of salt. In: International Symposium on Salts. Hamburg: Northern Ohio Geological Society; 1978

[35] Gorham P, Saltzberg D, Odian A, Williams D, Besson D, Frichter G, et al. Measurements of the suitability of large rock salt formations for radio detection of high energy neutrinos. Nuclear Instruments and Methods in Physics Research. 2002;**A490**:476-491

[36] Brancus I. s.a, Raport ştiinţific şi tehnic, IFIN-HH etapa 2, Pregătiri teoretice pentru detecţia radiaţiei cosmice în subteran, Proiect DETCOS nr. 82-104/2008

[37] Şteţ D. Ecrane electromagnetice, Curs 5 (1/2). Universitatea tehnică din Cluj Napoca. Cluj-Napoca, Romania: U.T. Press; 2011

[38] Available from: http://www.scritube.com/stiinta/fizica/Propagarea-undelor-electromagn42125.php

[39] Crețu TI. Fizică - Curs Universitar. București: Editura Tehnică; 1996

[40] Annan A, Davis JL, Gendzwill D. Radar sounding in potash mines, Saskatchewan, Canada. Geophysics. 1988;**53**:1556-1564

[41] Available from: http://media0.wgz.ro/files/media0:4b51f7d6af096.pdf.upl/ET5b-%20Incalzirea%

Chapter 8

Sensitivity and Uncertainty Quantification of Neutronic Integral Data Using ENDF/B-VII.1 and JENDL-4.0 Evaluations

Mustapha Makhloul, H. Boukhal, T. El Bardouni, E. Chakir, M. Kaddour and S. Elouahdani

Abstract

Many integral neutronic parameters such as the effective multiplication factors (k_{eff}) are based on neutron reactions with matter through cross sections. However, these cross sections present uncertainties, of origin multiple, which reduce the safety margin of nuclear installations. In order to minimize these risks, a sensitivity analysis is necessary to indicate the rate of change of a reactor performance parameter compared to variations in cross sections. Thus, several critical benchmarks were taken from the International Handbook of Evaluated Criticality Safety Benchmark Experiments (IHECSBE), and their sensitivities and covariance matrix of the desired cross section were processed by MCNP6 and NJOY codes, respectively, in ENDF/B-VII.1 and JENDL-4.0 evaluations. The results obtained show that the 44 energy groups give the most varied sensitivity profiles than those given by others (15 and 33). In addition, we observed large uncertainties on the k_{eff} due to the H-1 and O-16 cross-sectional uncertainties (~200–1000 pcm) in ENDF/B -VII.1 and the U-235 cross section in JENDL-4.0; however, k_{eff}'s uncertainties due to the cross-sectional uncertainties of the U-238 are very small.

Keywords: k_{eff}, sensitivity, covariance matrix, uncertainty, MCNP6.1, NJOY, multigroup cross section

1. Introduction

Prediction of integral nuclear parameters requires a reliable nuclear database such as microscopic nuclear parameters, cross sections, covariance matrices, etc. Many previous works [1, 2] have proved that the capture cross section of the uranium 235 has an important effect on the criticality calculations [3, 4]. For example, the relative uncertainty of k_{eff} in BFS core due to the ^{235}U capture cross-sectional uncertainty is near 202 pcm [5].

In present study, the uncertainty prediction in the multiplication factors is based on the ENDF/B-VII.1 and JENDL-4.0 evaluations where MCNP6 [6] Monte Carlo code is used for the sensitivity and k_{eff} calculations and the NJOY99 [7] is applied to calculate the covariances in three energy group structures (15, 33 and 44) for the most abundant isotopes in the studied benchmarks (^{235}U, ^{238}U, ^{1}H, and ^{16}O). All benchmarks were taken from IHECSBE [8].

2. Study approach

2.1 Multigroup structure

In this article, the effect of the multigroup energy of neutrons on the sensitivity of multiplication factors was studied for three cases (15, 33, and 44 groups). The covariances for many cross sections are often presented in the evaluated data libraries (ENDF/B-VII.1 and JENDL-4.0). All files were processed by the NJOY99 code to calculate the multigroup of interest cross sections in the ENDF-6 format. The modules RECONR and BROADR were used before to reconstruct the cross sections (MF = 3) at room temperature 300°K. The GROUPR module was used to generate the desired data in the grouped-wise format gendf for the three presentations (15, 33, and 44 groups) to retain the characteristic structure in the cross sections between 10^{-5} eV and 20 MeV. The energetic structures were generated from the fine-group library for resonance nuclides, with different weight flux functions: fission Maxwellian (10 MeV–70 keV), 1/E (70 keV–0.125 eV), and thermal Maxwellian (0.125–10^{-5} eV). **Tables 1–3** below present the three energy group structures.

Figures below illustrate the comparison of the pointwise and multigroup representations for the $^{235.238}U$ cross sections (**Figures 1** and **2**).

Figures above present that the pointwise and multigroup cross sections are very close in the two evaluations ENDF/B-VII.1 and JENDL-4.0.

2.2 Covariance data of cross sections

It is necessary to process the multigroup covariance matrices for each energy group structure (15, 33, and 44). Thus, an appropriate input file for nuclear code NJOY was prepared using several modules as ERROR, GROUPER, and COVR [11–13] to process the ENDF file (MF = 33) and generate the multigroup covariance matrices for the desired cross sections. The following figures show a comparison of these covariance matrices in the two evaluations studied using the structures of 15, 33, and 44 energy groups.

Figure 3 shows the uncertainty and covariance for the ^{235}U elastic cross section in the energy region from 10^{-5} eV to 20 MeV. In this figure, we can see that the

Group number	Energy range (eV)	Group number	Energy range (eV)
1	1.0000E-05	9	2.4800E+04
2	1.1000E-01	10	6.7400E+04
3	5.4000E-01	11	1.8300E+05
4	4.0000E+00	12	4.9800E+05
5	2.2600E+01	13	1.3500E+06
6	4.5400E+02	14	2.2300E+06
7	2.0400E+03	15	6.0700E+06
8	9.1200E+03	16	1.9600E+07

Table 1.
15 Neutron energy group structure [9].

Group number	Upper energy (eV)	Group number	Upper energy (eV)	Group number	Upper energy (eV)
1	1.0000E-01	12	4.5400E+02	23	1.1100E+05
2	5.4000E-01	13	7.4900E+02	24	1.8300E+05
3	4.0000E+00	14	1.2300E+03	25	3.0200E+05
4	8.3200E+00	15	2.0300E+03	26	4.9800E+05
5	1.3700E+01	16	3.3500E+03	27	8.2100E+05
6	2.2600E+01	17	5.5300E+03	28	1.3500E+06
7	4.0200E+01	18	9.1200E+03	29	2.2300E+06
8	6.7900E+01	19	1.5000E+04	30	3.6800E+06
9	9.1700E+01	20	2.4800E+04	31	6.0700E+06
10	1.4900E+02	21	4.0900E+04	32	1.0000E+07

Table 2.
33 Neutron energy group structure [10].

Group number	Upper energy (eV)	Group number	Upper energy (eV)	Group number	Upper energy (eV)
1	1.0000E-05	16	3.2500E-01	31	3.0000E+03
2	3.0000E-03	17	3.5000E-01	32	1.7000E+04
3	7.5000E-03	18	3.7500E-01	33	2.5000E+04
4	1.0000E-02	19	4.0000E-01	34	1.0000E+05
5	2.5300E-02	20	6.2500E-01	35	4.0000E+05
6	3.0000E-02	21	1.0000E+00	36	9.0000E+05
7	4.0000E-02	22	1.7700E+00	37	1.4000E+06
8	5.0000E-02	23	3.0000E+00	38	1.8500E+06
9	7.0000E-02	24	4.7500E+00	39	2.3540E+06
10	1.0000E-01	25	6.0000E+00	40	2.4790E+06
11	1.5000E-01	26	8.1000E+00	41	3.0000E+06
12	2.0000E-01	27	1.0000E+01	42	4.8000E+06
13	2.2500E-01	28	3.0000E+01	43	6.4340E+06
14	2.5000E-01	29	1.0000E+02	44	8.1873E+06
15	2.7500E-01	30	5.5000E+02	45	2.0000E+07

Table 3.
44 Neutron energy group structure [10].

lowest uncertainty is given by the 44-group structure where around the energy 10 keV, the uncertainty is ~4% in the ENDF/B-VII.1 and ~ 9.5% in JENDL-4.0. In addition, negative correlations are observed in JENDL-4.0.

According to **Figure 4**, the maximum uncertainties in the fission cross sections of the ^{235}U in the energy less than 10 eV are, respectively, ~7.5% and ~1% in JENDL-4.0 and ENDF/B-VII.1 for 15 and 33 groups; however, in the 44-group structure, one can see that this maximum is ~15% around the energy 3 eV. In the energy interval [10 eV; 20 MeV], these uncertainties are very close to 1% for the two evaluations in 33- and 44-group structures, while for the 15-group structure,

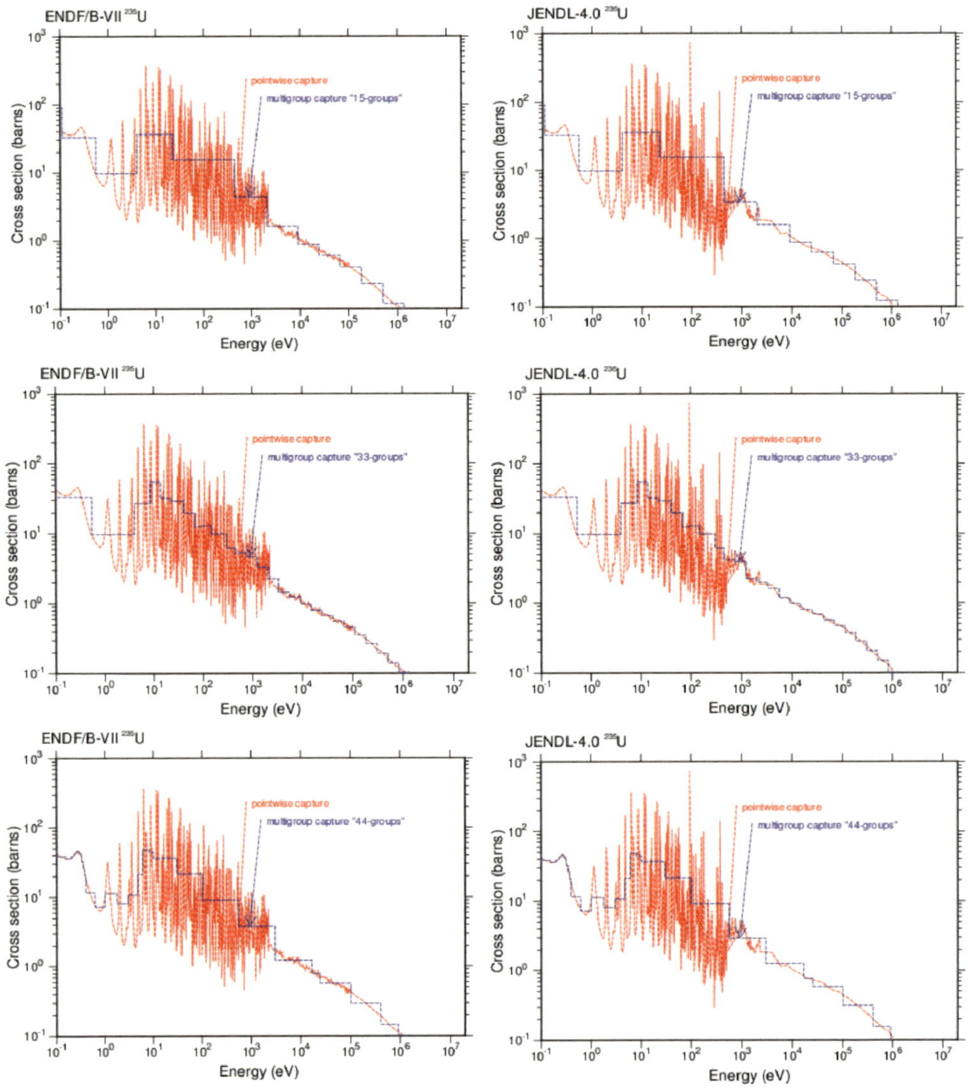

Figure 1.
The pointwise and multigroup (15, 33, and 44) capture cross section for the ^{235}U.

the uncertainties in JENDL-4.0 are higher than those in ENDF/B-VII.1. Also, negative correlations appeared in JENDL-4.0.

2.3 Sensitivity-uncertainty theory

Sensitivity coefficients represent the percentage effect on some nuclear system response (e.g., multiplication factor k_{eff}) due to a percentage change in an input parameter such as cross section (capture, fission, elastic, inelastic, etc.). The sensitivity of k_{eff} (noted simply k) to a multigroup cross section $\sigma_{x.g}$, for an energy group g, is defined according to [14] by Eq. (1), where the first order of the perturbation theory is used [15–17]:

$$S_{x.g} = \frac{\sigma_{x.g}}{k} \frac{\partial k}{\partial \sigma_{x.g}} \tag{1}$$

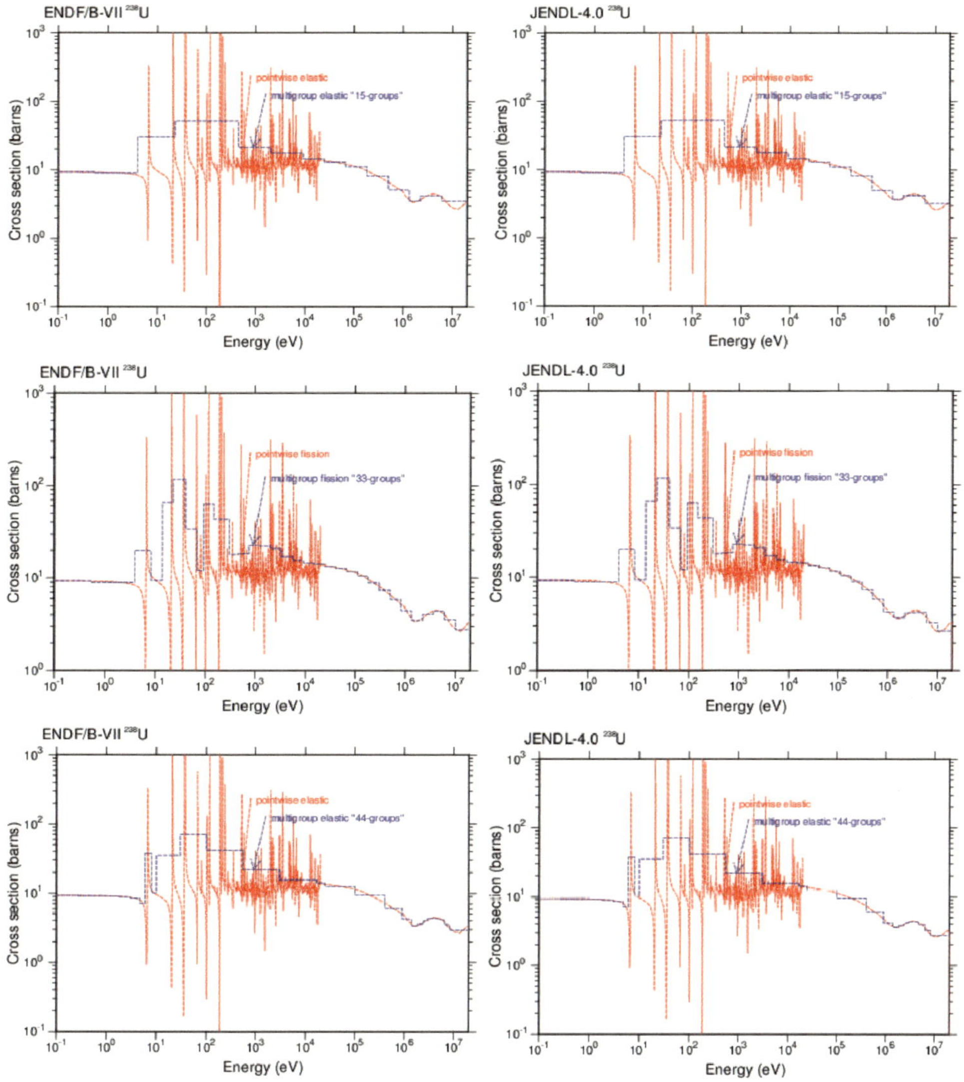

Figure 2.
The pointwise and multigroup (15, 33, and 44) elastic cross section for the ^{238}U.

These coefficients are supposed to be constant in the first order of perturbation theory, so the sensitivity matrix S (Eq. (2)) is also constant [18–20]:

$$S_k = \left[\frac{\sigma_i}{k_j}\frac{\partial k_j}{\partial \sigma_j}\right] \; i = 1.2. \ldots .n \; and \; j = 1.2. \ldots .m \tag{2}$$

where m is the number of critical systems considered and n is the number of energy groups (n = 15, 33, and 44).

The sensitivity matrix coefficients have been calculated using MCNP6.1 code using KSEN card.

The integral quantities calculated with a reference cross section set σ are denoted by k. The integral quantities k’ calculated with a cross section set σ′, which deviates by $\delta\sigma$ from σ, have the following relation with k:

$$k' = k(1 + S.\delta\sigma) \tag{3}$$

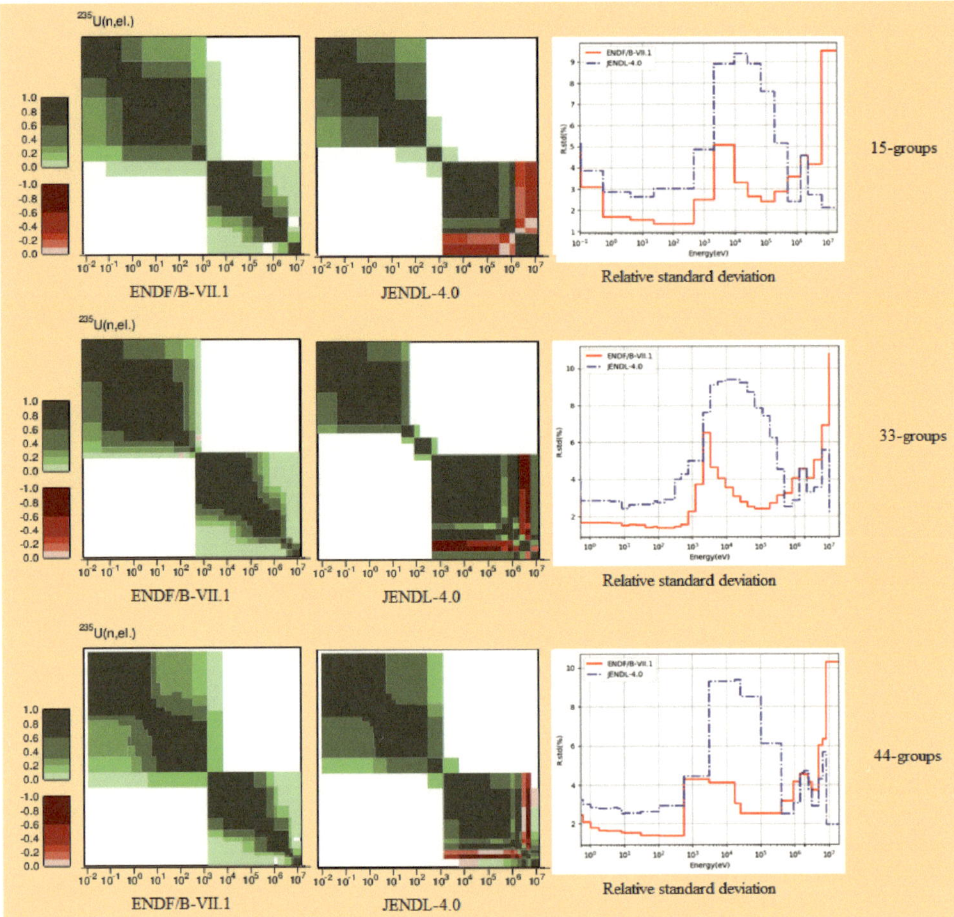

Figure 3.
Uncertainty and covariance for the 235 U elastic cross section using 15, 33, and 44 structure energy groups in ENDF/B-VII.1 and JENDL-4.0 evaluations.

The covariance of k'/k is given by

$$V_k = (S\sigma' - S\sigma)(S\sigma' - S\sigma)^t$$
$$V_k = S(T - T_0)(T - T_0)^t S^t$$
$$V_k = SMS^t \tag{4}$$

where t stands for the transpose of the matrix S.

The square root of the diagonal term V_{ii} of V_k is the standard deviation in the integral quantity k_i. Thus, the prior nuclear data uncertainty of k can be obtained in matrix expression form by the so-called sandwich rule [20, 21]:

$$(\Delta k)^2 = SMS^t \tag{5}$$

The non-diagonal term $V_{ij}(i \neq j)$ gives the degree of correlation between the errors of k_i and k_j. The element r_{ij} of the correlation matrix is obtained by dividing the element V_{ij} by the products of standard deviation V_{ii} and V_{jj}:

$$r_{ij} = \frac{V_{ij}}{\sqrt{V_{ii}V_{jj}}} \tag{6}$$

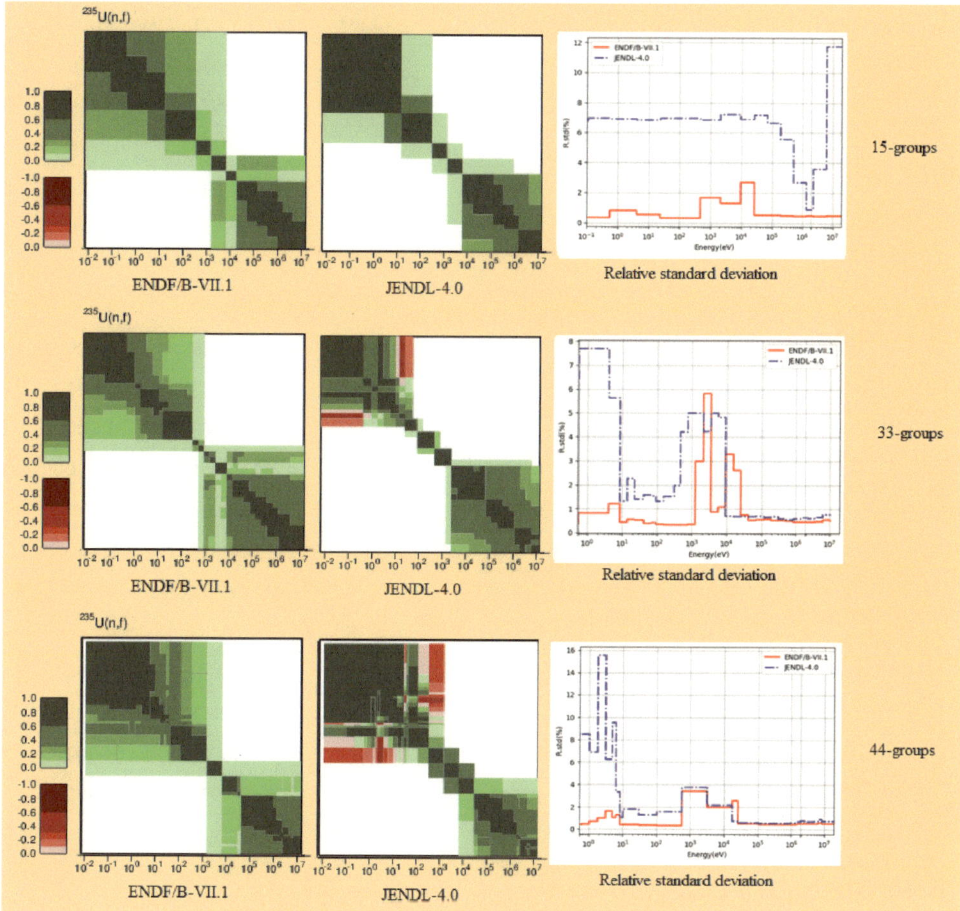

Figure 4.
Uncertainty and covariance for the ^{235}U fission cross section using 15, 33, and 44 energy group structures in ENDF/B-VII.1 and JENDL-4.0 evaluations.

A common practice in uncertainty calculations is the relative sensitivity coefficients provided from the sensitivity analysis. Therefore, the relative matrices are used as

$$(\Delta k/k)^2 = GPG^t \tag{7}$$

where G is the relative sensitivity matrix and P is the relative covariance matrix of the interest cross section.

Eq. (7) mathematically links the uncertainty of the integral data and the uncertainties of the cross sections through the associated sensitivity coefficients. Thus, a high sensitivity and an uncertain cross section generate a large uncertainty in k.

3. Results and discussion

3.1 Description of benchmark systems

In this work, 25 HEU-SOL-THERM thermal experiments, 4 HEU-MET-INTER intermediate experiments, and 21 HEU-MET-FAST fast experiments are studied. The calculated, experimental k_{eff} and their uncertainties for each benchmark are

Benchmarks	k.exp	std.exp	k.cal (ENDF/B-VII.1)	std.cal	k.cal (JENDL-4.0)	std.cal
Hst001.001	1.0004	0.0060	0.99815	0.00005	0.99952	0.00005
Hst001.002	1.0021	0.0072	0.99722	0.00006	0.99927	0.00006
Hst001.003	1.0003	0.0035	1.00155	0.00005	1.00296	0.00005
Hst001.004	1.0008	0.0053	0.99815	0.00005	1.00048	0.00005
Hst001.005	1.0001	0.0049	0.99874	0.00005	0.99949	0.00005
Hst001.006	1.0002	0.0046	1.00196	0.00005	1.00283	0.00005
Hst001.007	1.0008	0.0040	0.99781	0.00005	0.99910	0.00005
Hst001.008	0.9998	0.0038	0.99797	0.00005	0.99930	0.00005
Hst001.009	1.0008	0.0054	0.99412	0.00006	0.99633	0.00006
Hst001.010	0.9993	0.0054	0.99241	0.00005	0.99338	0.00005
Hst009.001	0.9990	0.0043	0.99695	0.00005	1.00096	0.00005
Hst009.002	1.0000	0.0039	0.99686	0.00005	1.00034	0.00005
Hst009.003	1.0000	0.0036	0.99556	0.00005	0.99830	0.00005
Hst009.004	0.9986	0.0035	0.98894	0.00005	0.99112	0.00005
Hst009.010	1.0000	0.0057	0.99745	0.00005	1.00153	0.00005
Hst010.001	1.0000	0.0029	0.99453	0.00005	0.99633	0.00005
Hst010.002	1.0000	0.0018	0.99496	0.00005	0.99678	0.00005
Hst010.003	1.0000	0.0029	0.99247	0.00005	0.99237	0.00005
Hst010.004	0.9992	0.0029	0.99052	0.00005	0.99994	0.00004
Hst011.001	1.0000	0.0023	0.99859	0.00004	0.99773	0.00003
Hst011.002	1.0000	0.0023	0.99866	0.00004	0.99602	0.00004
Hst012.001	0.9999	0.0058	0.99723	0.00003	0.99745	0.00004
Hst013.001	1.0012	0.0026	0.99868	0.00003	1.00569	0.00005
Hst028.001	1.0000	0.0023	0.99642	0.00005	0.99580	0.00004
Hst035.007	1.0000	0.0035	1.00467	0.00005	0.99938	0.00004
Hmi006.001	0.9977	0.0008	0.99297	0.00004	1.00151	0.00004
Hmi006.002	1.0001	0.0008	0.99682	0.00004	1.00315	0.00004
Hmi006.003	1.0015	0.0009	1.00082	0.00004	0.99751	0.00003
Hmi006.004	1.0016	0.0008	1.00732	0.00004	0.99025	0.00003
Hmf001.001	1.0004	0.0024	0.99976	0.00003	0.98929	0.00003
Hmf003.001	1.0000	0.0050	0.99501	0.00003	0.99386	0.00003
Hmf003.002	1.0000	0.0050	0.99436	0.00003	0.99208	0.00003
Hmf003.003	1.0000	0.0050	0.99918	0.00003	0.99638	0.00003
Hmf003.004	1.0000	0.0050	0.99721	0.00003	1.00187	0.00003
Hmf003.005	1.0000	0.0030	1.00146	0.00003	1.00190	0.00003
Hmf003.008	1.0000	0.0030	1.00214	0.00003	1.00515	0.00003
Hmf003.009	1.0000	0.0050	1.00244	0.00003	1.00960	0.00003
hmf003.010	1.0000	0.0050	1.00505	0.00003	0.99315	0.00003
Hmf003.011	1.0000	0.0030	1.00886	0.00003	0.99656	0.00004
Hmf008.001	0.9989	0.0016	0.99577	0.00003	0.99518	0.00003

Benchmarks	k.exp	std.exp	k.cal (ENDF/B-VII.1)	std.cal	k.cal (JENDL-4.0)	std.cal
Hmf011.001	0.9989	0.0015	0.99887	0.00004	0.99265	0.00003
Hmf012.001	0.9992	0.0018	0.99810	0.00003	0.99236	0.00003
Hmf014.001	0.9989	0.0017	0.99774	0.00003	0.99684	0.00003
Hmf015.001	0.9996	0.0017	0.99447	0.00003	0.99774	0.00003
Hmf018.002	1.0000	0.0014	0.99946	0.00003	0.99337	0.00003
Hmf020.002	1.0000	0.0028	1.00057	0.00003	0.99416	0.00003
Hmf021.002	1.0000	0.0024	0.99750	0.00003	1.00132	0.00004
Hmf022.002	1.0000	0.0021	0.99746	0.0003	0.99782	0.00003
Hmf026.011	0.9982	0.0042	1.00312	0.00004	1.00647	0.00005
Hmf028.001	1.0000	0.0030	1.00286	0.00003	1.00745	0.00005

Table 4.
K_{eff} *benchmark cases and their statistical uncertainties (1σ).*

	Hst001.001		Hst001.006		Hst035.007	
Isotope	**ENDF/B-VII.1**	**JENDL-4.0**	**ENDF/B-VII.1**	**JENDL-4.0**	**ENDF/B-VII.1**	**JENDL-4.0**
U-234	-2.7926E-03	-2.6083E-03	-2.0123E-03	-1.8550E-03	-2.3645E-03	-2.4621E-03
U-235	1.0503E-01	1.1931E-01	2.2592E-01	2.2535E-01	1.3115E-01	1.3959E-01
U-236	-2.9683E-04	-4.2700E-04	-5.0385E-05	-1.6201E-04	-2.3107E-04	-2.4627E-04
U-238	-2.7572E-03	-3.3237E-03	-1.1858E-03	-1.3662E-03	-6.4877E-03	-5.4229E-03
H-1	5.5723E-01	5.4111E-01	3.4059E-01	3.4260E-01	3.9636E-01	4.0646E-01
O-16	1.3385E-01	1.3190E-01	1.1264E-01	1.1672E-01	9.0869E-02	9.3703E-02
N-14	-2.8936E-03	-6.3375E-04	-2.9080E-03	-5.5300E-03	-3.6365E-03	-5.4211E-03

Table 5.
Total integrated sensitivity for thermal benchmark (%%).

	Hmf001.001		Hmf003.001	
Isotope	**ENDF/B-VII.1**	**JENDL-4.0**	**ENDF/B-VII.1**	**JENDL-4.0**
U-234	5.1587E-03	7.4997E-03	1.4957E-03	1.9720E-03
U-235	8.0536E-01	8.0423E-01	1.4577E-01	1.5009E-01
U-238	1.7438E-02	1.7762E-02	4.2719E-02	4.1731E-02

Table 6.
Total integrated sensitivity for fast benchmark (%%).

summarized in **Table 4**. All calculations were performed using 100,000 neutrons per cycle, 150 inactive cycles, and 4000 active cycles to minimize statistical uncertainty (~5 pcm).

3.2 Total sensitivity evaluation

The total sensitivity calculations were performed in order to identify the most important cross sections for neutron-induced reactions in critical experiments summarized in **Table 4**. The total integrated sensitivities obtained using the ENDF/B-VII.1 and JENDL-4.0 evaluations are presented in **Tables 5** and **6**. We can see from

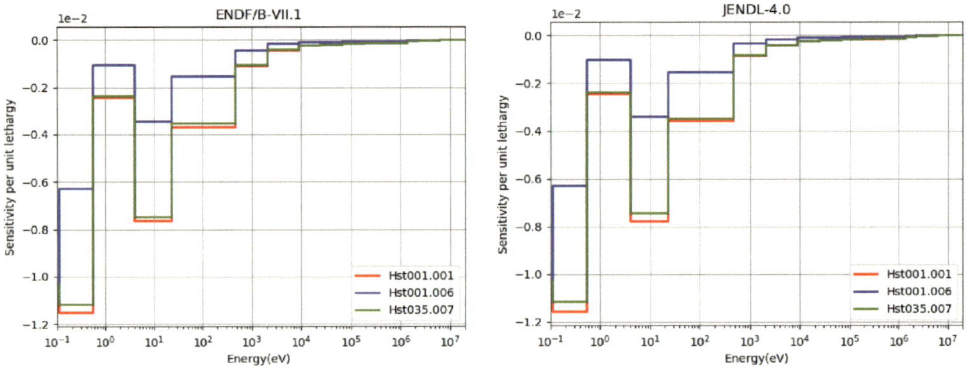

Figure 5.
Sensitivity profiles of ^{235}U capture cross section for thermal benchmarks with 15 energy groups—ENDF/B-VII.1 and JENDL-4.0.

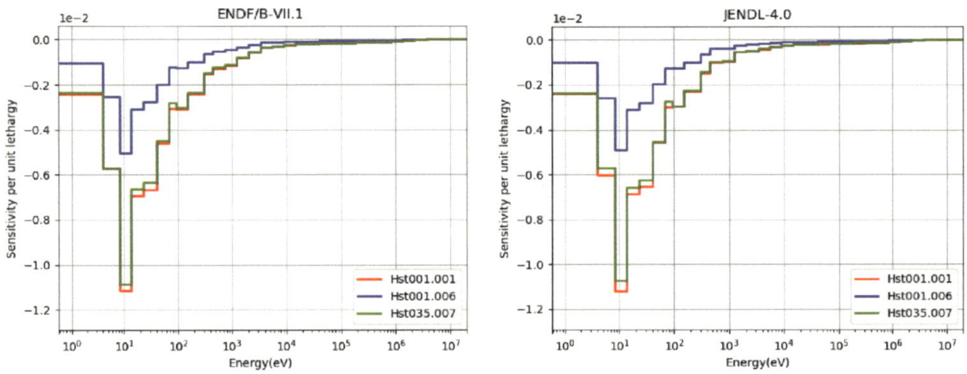

Figure 6.
Sensitivity profiles of ^{235}U capture cross section for thermal benchmarks with 33 energy groups—ENDF/B-VII.1 and JENDL-4.0.

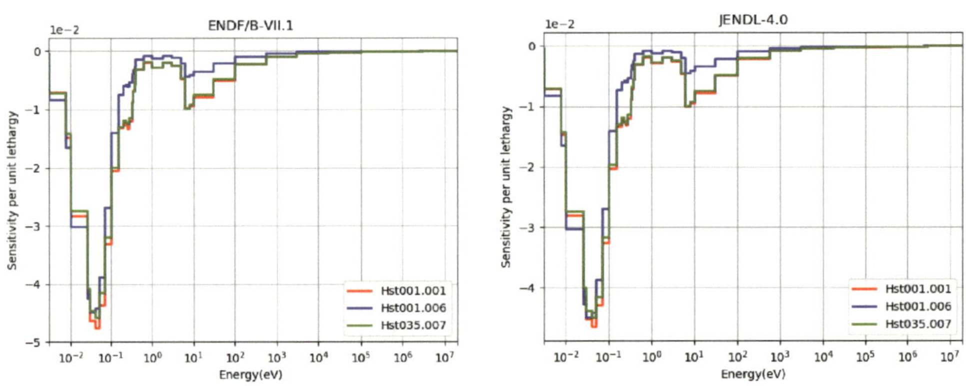

Figure 7.
Sensitivity profiles of ^{235}U capture cross section for thermal benchmarks with 44 energy groups—ENDF/B-VII.1 and JENDL-4.0.

these tables that the total integrated sensitivity obtained with the two nuclear evaluations is almost the same; in addition, the sensitivities of U-234, U-236, and N-14 are very low compared to the others. Thus, for the quantification of sensitivity and uncertainty, only U-235, U-238, H-1, and O-16 are taken into account.

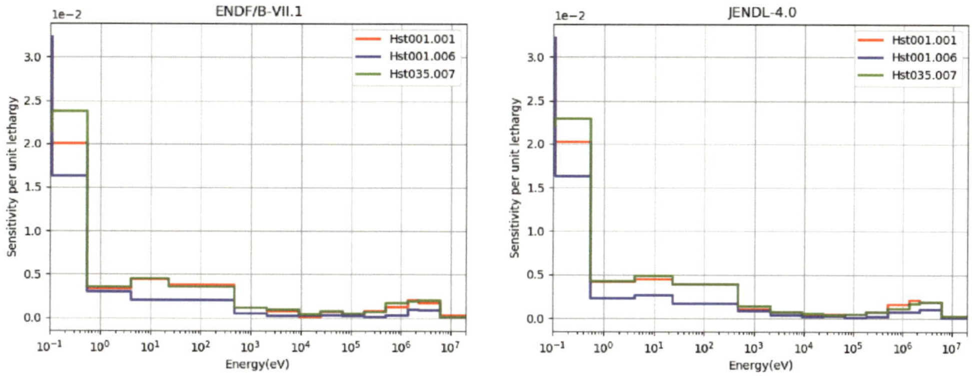

Figure 8.
Sensitivity profiles of ^{235}U fission cross section for thermal benchmarks with 15 energy groups—ENDF/B-VII.1 and JENDL-4.0.

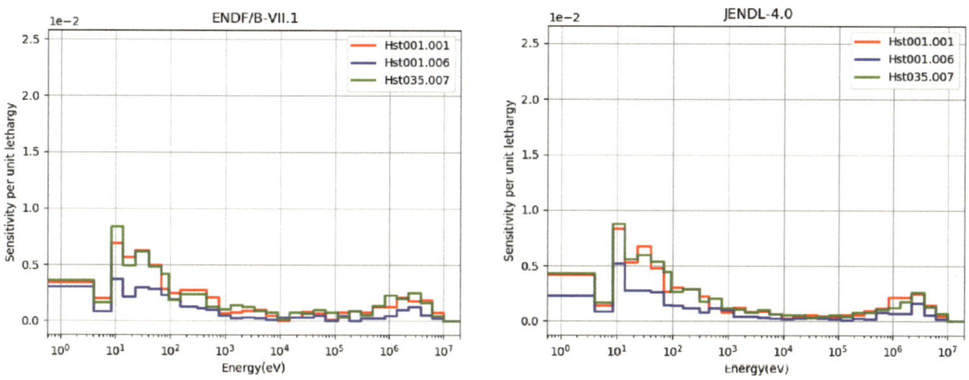

Figure 9.
Sensitivity profiles of ^{235}U fission cross section for thermal benchmarks with 33 energy groups—ENDF/B-VII.1 and JENDL-4.0.

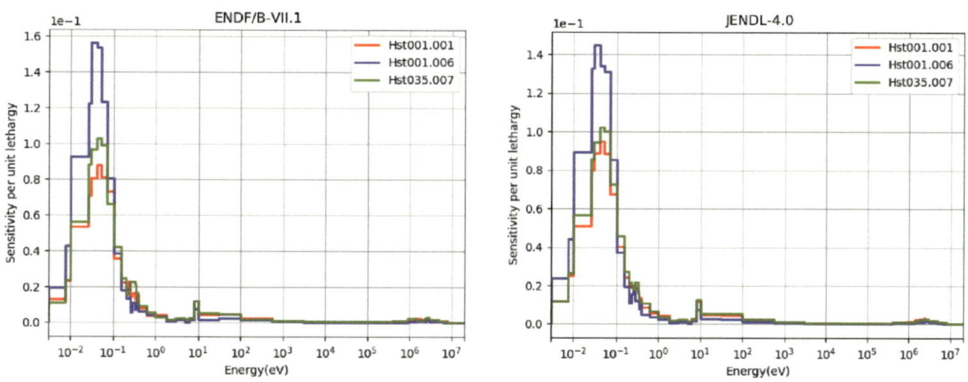

Figure 10.
Sensitivity profiles of ^{235}U fission cross section for thermal benchmarks with 44 energy groups—ENDF/B-VII.1 and JENDL-4.0.

3.3 Sensitivities of k_{eff} with respect to multigroup cross section

In this study, the sensitivity coefficients obtained with the two libraries ENDF/B-VII.1 and JENDL-4.0 are evaluated using MCNP6.1 code in three

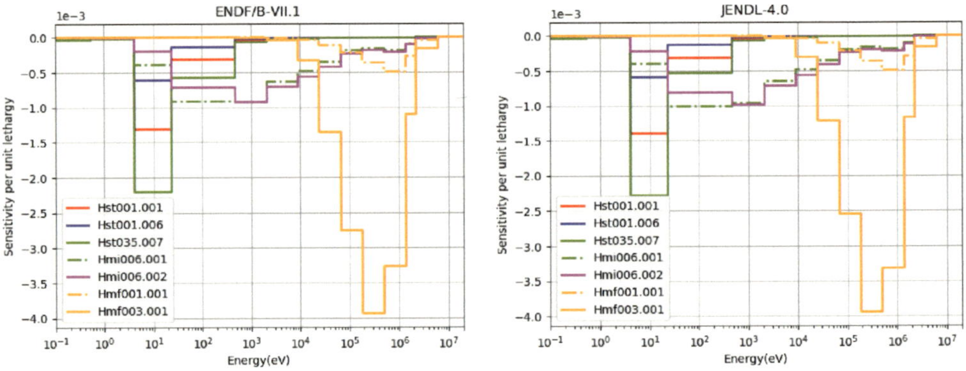

Figure 11.
Sensitivity profiles of the ^{238}U capture cross section with 15 energy groups—ENDF/B-VII.1 and JENDL-4.0.

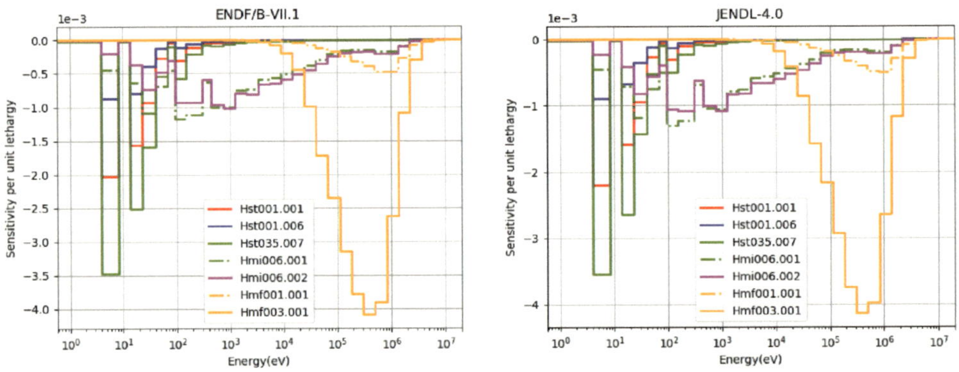

Figure 12.
Sensitivity profiles of the ^{238}U capture cross section with 33 energy groups—ENDF/B-VII.1 and JENDL-4.0.

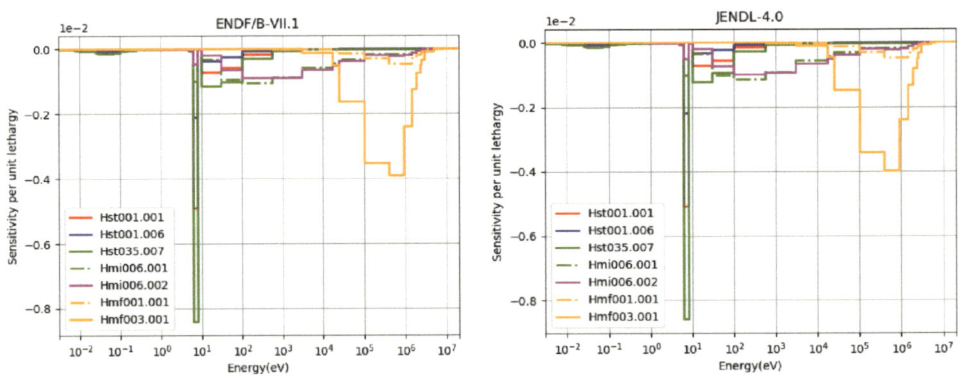

Figure 13.
Sensitivity profiles of the ^{238}U capture cross section with 44 energy groups—ENDF/B-VII.1 and JENDL-4.0.

multigroup structures (15, 33, and 44). Given the large number of cross sections for each energy group, the sensitivities of certain cross sections are only presented.

3.3.1 Sensitivity for the ^{235}U cross section

The results obtained are presented in the figures below for the ^{235}U cross sections (**Figures 5–10**).

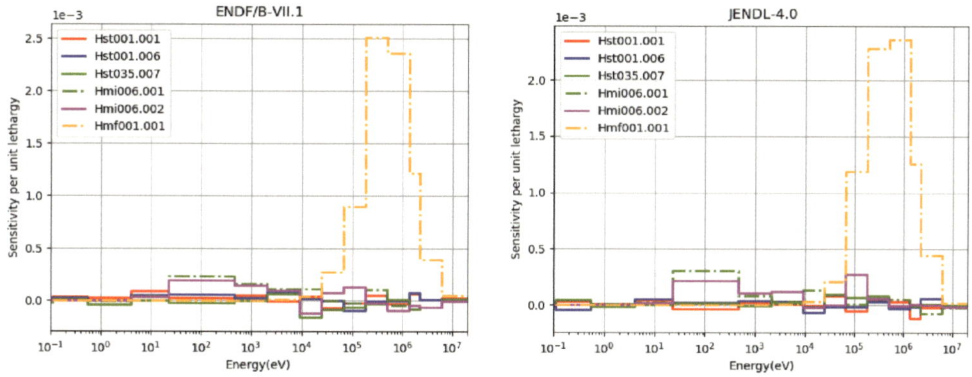

Figure 14.
Sensitivity profiles of the ^{238}U elastic cross section with 15 energy groups—ENDF/B-VII.1 and JENDL-4.0.

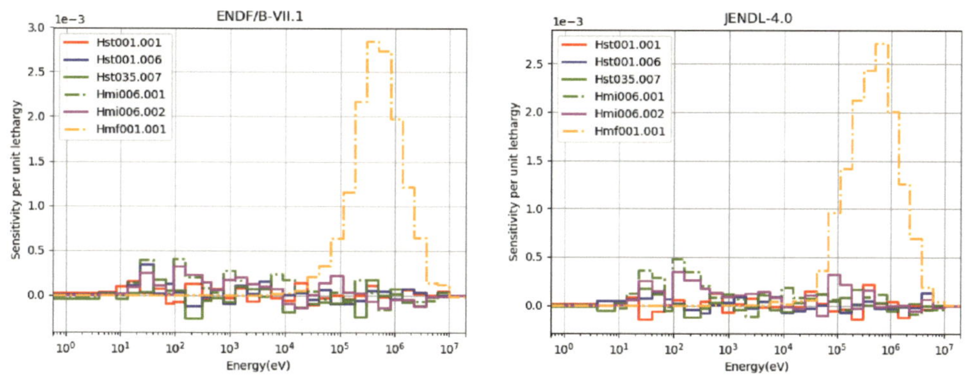

Figure 15.
Sensitivity profiles of the ^{238}U elastic cross section with 33 energy groups—ENDF and JENDL-4.0.

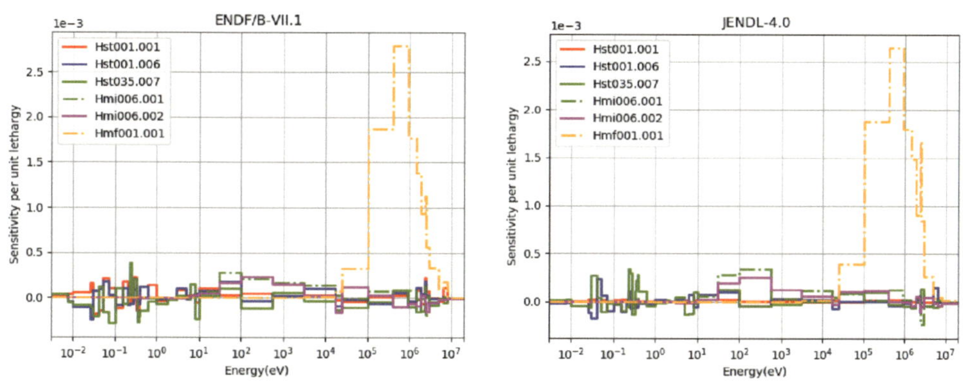

Figure 16.
Sensitivity profiles of the ^{238}U elastic cross section with 44 energy groups—ENDF and JENDL-4.0.

From these figures, it can be seen that the 44-group structure of energy gives very varied sensitivity profiles depending on the neutron energy; moreover, this group structure gives sensitivities slightly lower than those given by the structures of 15 and 33 groups. Consequently, it can be said that the precision of the sensitivity increases for the structure containing the largest number of energy groups. Thus, low uncertainties on nuclear data are expected with this structure (44 groups) in the two evaluations (ENDF/B-VII.1 and JENDL-4.0).

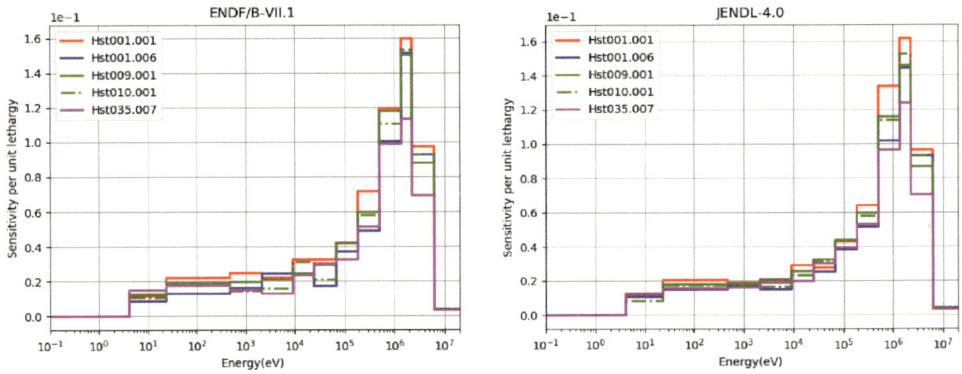

Figure 17.
Sensitivity profiles of the ^{1}H elastic cross section with 15 energy groups—ENDF/B-VII.1 and JENDL-4.0.

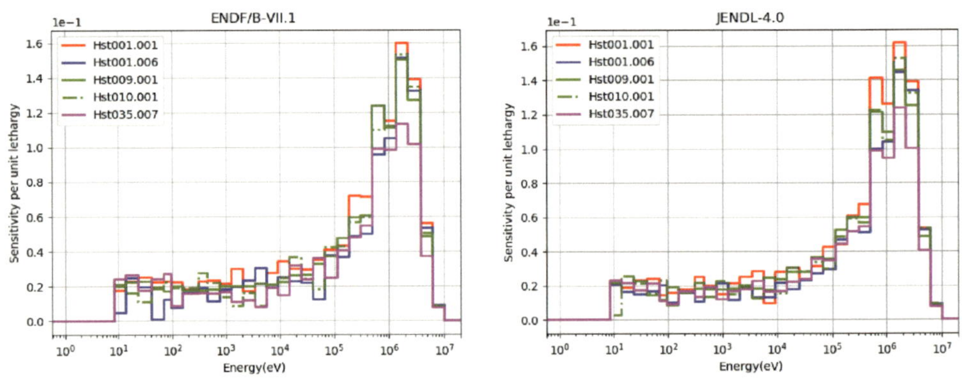

Figure 18.
Sensitivity profiles of the ^{1}H elastic cross section with 33 energy groups—ENDF/B-VII.1 and JENDL-4.0.

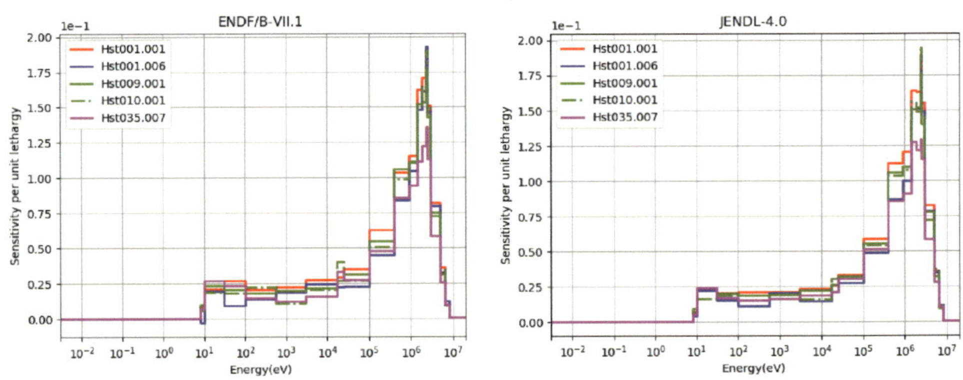

Figure 19.
Sensitivity profiles of the ^{1}H elastic cross section with 44 energy groups—ENDF/B-VII.1 and JENDL-4.0.

3.3.2 Sensitivity for the ^{238}U cross section

The sensitivities of the multiplication factors for the ^{238}U cross sections are shown in the figures below (**Figures 11–16**).

These figures show that thermal and intermediate critical experiment designs demonstrate low sensitivity to the capture and elastic cross sections of the ^{238}U at

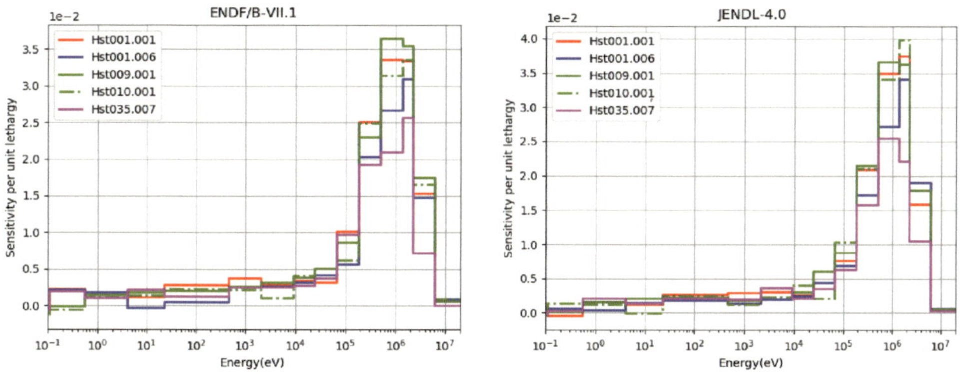

Figure 20.
Sensitivity profiles of the ^{16}O elastic cross section with 15 energy groups—ENDF/B-VII.1 and JENDL-4.0.

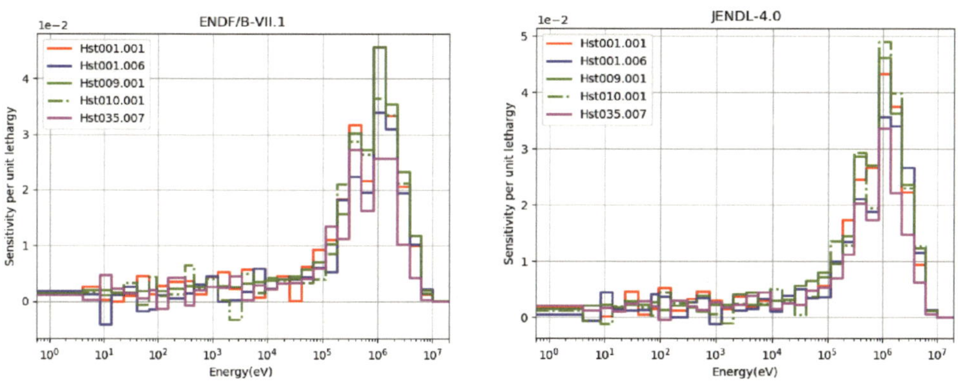

Figure 21.
Sensitivity profiles of the ^{16}O elastic cross section with 33 energy groups—ENDF/B-VII.1 and JENDL-4.0.

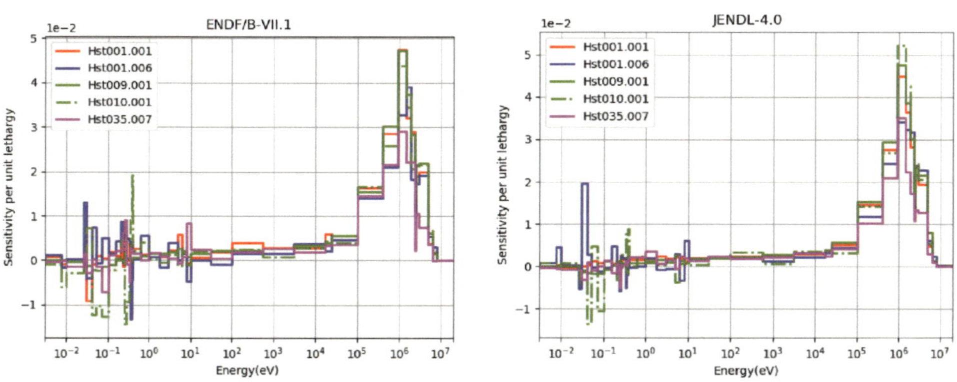

Figure 22.
Sensitivity profiles of the ^{16}O elastic cross section with 44 energy groups—ENDF/B-VII.1 and JENDL-4.0.

high energies and a significant sensitivity at thermal and resonance energies. However, the fast critical experiments demonstrate, at high energies, high levels of sensitivity to the capture and elastic cross sections of the ^{238}U. Also, the structure of the 44 energy groups gives very varied sensitivity profiles compared to those given by structures 15 and 33.

3.3.3 *Sensitivity for ^{1}H and ^{16}O cross sections*

The sensitivities of the k_{eff}'s with respect to the cross sections of ^{1}H and ^{16}O are presented in the figures below (**Figures 17–22**).

The figures above show that all thermal critical benchmarks demonstrate low sensitivity to the ^{1}H and ^{16}O elastic cross sections for low resonance energies and significant sensitivity for high energies. In addition, the structure of the 44 energy groups gives very varied sensitivity profiles compared to those given by the 15- and 33-group structures. Also, the sensitivities given by ENDF/B-VII.1 are slightly lower than those given by JENDL-4.0.

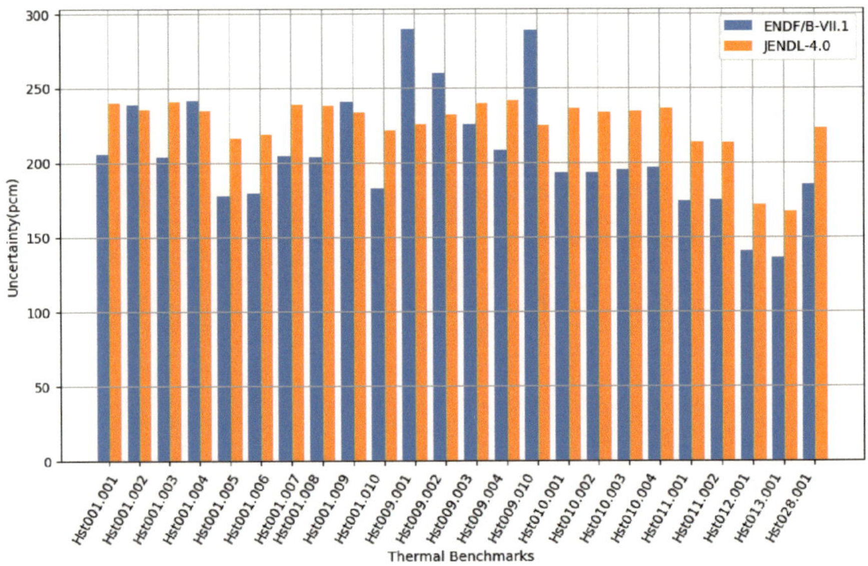

Figure 23.
Δk/k (pcm) prediction due to the uncertainties in ^{235}U capture cross sections with 44 energy groups for thermal benchmarks—ENDF/B-VII.1 and JENDL-4.0.

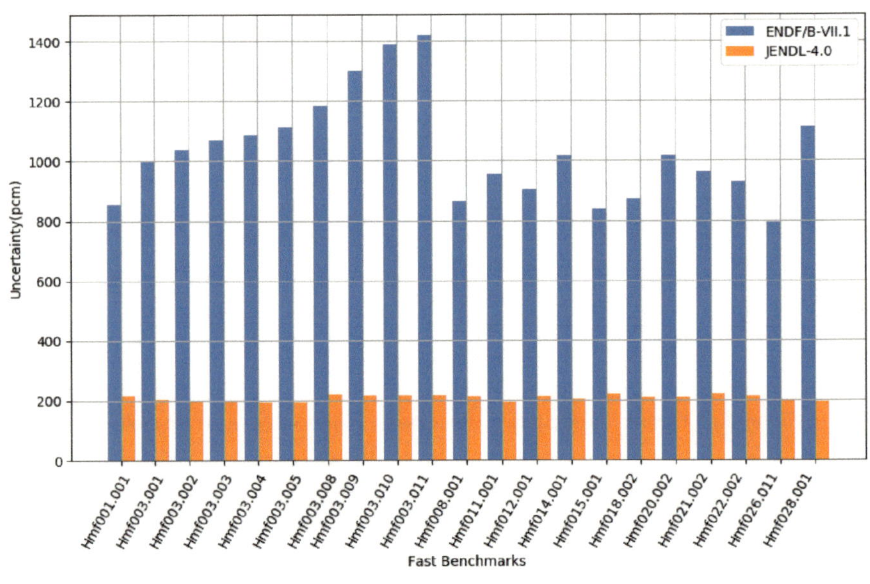

Figure 24.
Δk/k (pcm) prediction due to the uncertainties in ^{235}U capture cross sections with 44 energy groups for fast benchmarks—ENDF/B-VII.1 and JENDL-4.0.

In the following, all the results concerning only the structure of the group of 44 neutrons and presented for both ENDF/B-VII.1 and JENDL-4.0.

3.4 Nuclear data uncertainty prediction of k_{eff}

The nuclear data uncertainties of k_{eff} are calculated using Eq. (7), and the predictions of $\Delta k/k$ due to the uncertainty of the ^{235}U cross sections are presented in the figures below (**Figures 23–26**).

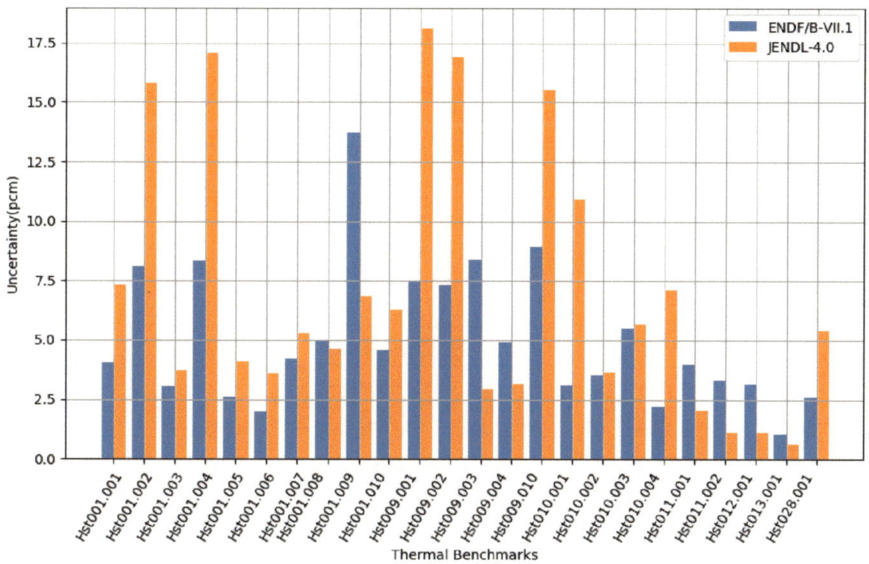

Figure 25.
Δk/k (pcm) prediction due to the uncertainties in ^{235}U elastic cross sections with 44 energy groups for thermal benchmarks—ENDF/B-VII.1 and JENDL-4.0.

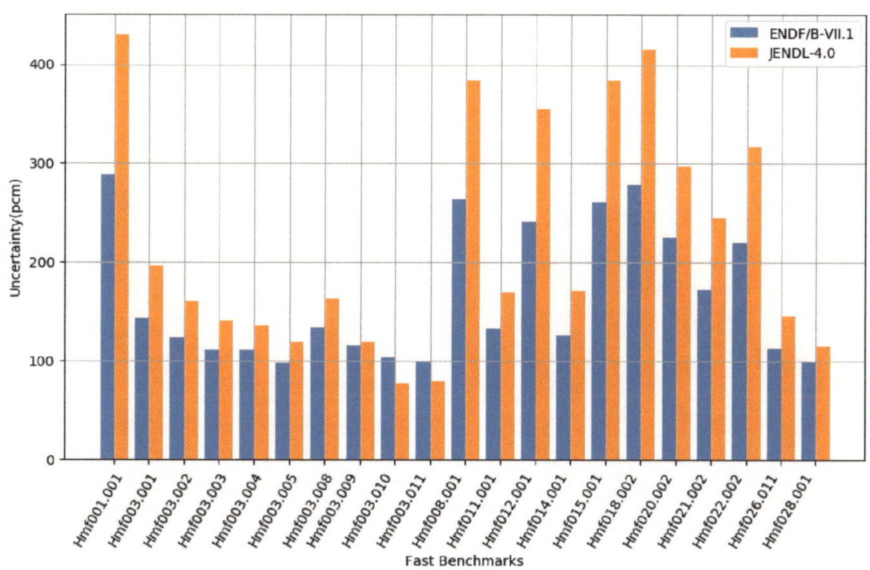

Figure 26.
Δk/k (pcm) prediction due to the uncertainties in ^{235}U elastic cross sections with 44 energy groups for fast benchmarks—ENDF/B-VII.1 and JENDL-4.0.

Thermal benchmarks	$\Delta k_{eff}/k_{eff}$-(%) ENDF/B-VII.1	$\Delta k_{eff}/k_{eff}$-(%) JENDL-4.0	Fast benchmarks	$\Delta k_{eff}/k_{eff}$-(%) ENDF/B-VII.1	$\Delta k_{eff}/k_{eff}$-(%) JENDL-4.0
Hst001.001	0.962	1.564	Hmi006.001	0.617	0.389
Hst001.002	0.931	1.214	Hmi006.002	0.800	0.274
Hst001.003	0.963	1.566	Hmi006.003	0.628	0.143
Hst001.004	0.934	1.200	Hmi006.004	0.827	0.159
Hst001.005	1.021	2.219	Hmf001.001	0.439	0.063
Hst001.006	1.011	2.160	Hmf003.001	0.442	0.057
Hst001.007	0.965	1.609	Hmf003.002	0.490	0.075
Hst001.008	0.963	1.587	Hmf003.003	0.492	0.083
Hst001.009	0.940	1.227	Hmf003.004	0.598	0.129
Hst001.010	0.995	2.104	Hmf003.005	0.446	0.053
Hst009.001	0.918	1.051	Hmf003.008	0.702	0.189
Hst009.002	0.929	1.122	Hmf003.009	0.667	0.159
Hst009.003	0.947	1.281	hmf003.010	0.603	0.141
Hst009.004	0.980	1.471	Hmf003.011	0.543	0.096
Hst009.010	0.917	1.065	Hmf008.001	0.648	0.221
Hst010.001	1.004	1.785	Hmf011.001	0.505	0.999
Hst010.002	1.009	1.820	Hmf012.001	0.458	0.086
Hst010.003	1.004	1.800	Hmf014.001	0.462	0.070
Hst010.004	1.010	1.747	Hmf015.001	0.463	0.084
Hst011.001	1.069	2.295	Hmf018.002	0.437	0.057
Hst011.002	1.074	2.306	Hmf020.002	0.447	0.244
Hst012.001	1.274	3.150	Hmf021.002	0.423	0.035
Hst013.001	1.281	3.237	Hmf022.002	0.421	0.035
Hst028.001	1.007	2.066	Hmf026.011	0.505	1.326
Hst035.007	0.880	1.682	Hmf028.001	0.489	0.075

Table 7.
Total uncertainty of k_{eff} (%).

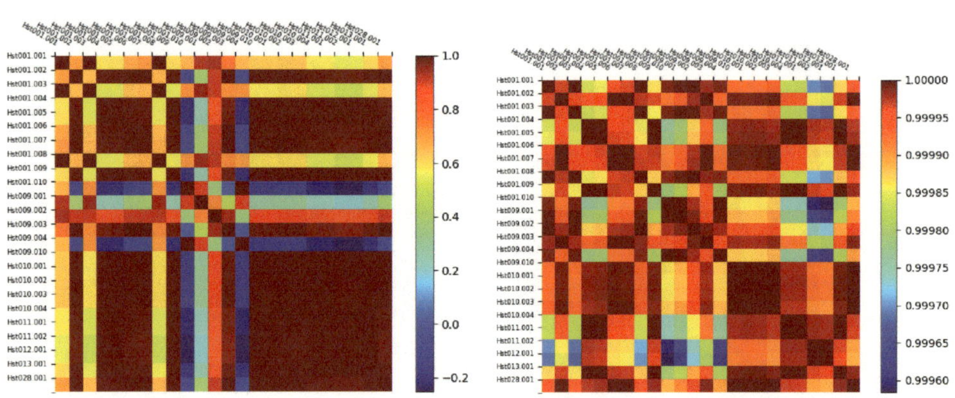

(a) (b)

Figure 27.
Correlation between thermal benchmarks due to the uncertainties in ^{235}U capture cross sections. (a) ENDF/B-VII.1 (b) JENDL-4.0.

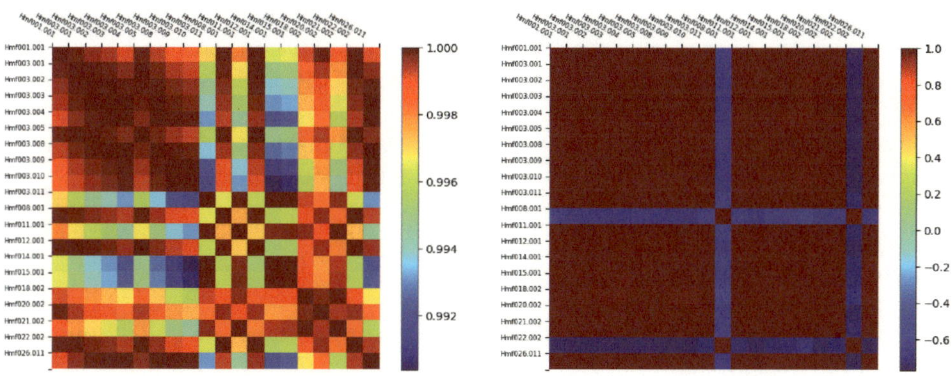

(a) **(b)**

Figure 28.
Correlation between fast benchmarks due to the uncertainties in ^{1}H and ^{16}O capture and elastic cross sections. (a) ENDF/B-VII.1 (b) JENDL-4.0.

We can see from these figures that, in general, the relative uncertainties of keff using ENDF/B-VII.1 are less than those using JENDL-4.0. For example, these uncertainties due to the ^{235}U capture cross section are ~250 pcm, and in the elastic cross section case, they are ~15 pcm for thermal benchmarks and 100–400 pcm for fast benchmarks. Concerning the predictions $\Delta k/k$ due to the uncertainty of the ^{238}U cross sections, they are all very small except for the elastic and inelastic cross sections.

The total uncertainties of the effective multiplication factors due to U-235, U-238, H-1, and O-16 are summarized in **Table 7**.

Table 7 shows that for thermal benchmarks, the relative uncertainties of k_{eff} with the ENDF/B-VII.1 are lower than those with JENDL-4.0. However, for fast benchmarks, these uncertainties are greater with the JENDL-4.0 evaluation.

3.5 Correlation between benchmark errors

The degree of correlation between benchmarks errors is calculated using Eq. (6); the results obtained are presented in the figures below.

Figures 27 and **28** show that the correlations between the benchmarks using ENDFB-VII.1 are lower than those using JENDL-4.0. Thus, a close similarity between several experiences is noted.

4. Conclusions

The multigroup effect on the sensitivities of k_{eff} with respect to cross sections U-235, U-238, H-1, and O-16 is studied using 15, 33, and 44 energy groups. We found that the structure of the 44 groups gives the most varied sensitivity profiles in the two evaluations ENDF/B-VII.1 and JENDL-4.0, allowing a better investigation of the uncertainties of the nuclear data.

The results obtained show that the k_{eff} sensitivity profiles are approximately the same for the two nuclear evaluations ENDF/B-VII.1 and JENDL-4.0. However, the covariances of the cross sections are different between the two evaluations, which is why differences between the uncertainties of the nuclear data are observed between these evaluations. For example, the total uncertainties in the thermal benchmark Hst001.001 are, respectively, 0.962 and 1.564% with ENDF/B-VII.1 and JENDL-4.0, and for the fast benchmark Hmf.001.001, these uncertainties are 0.439 and

0.063% with ENDF/B -VII.1 and JENDL-4.0, respectively. These differences are mainly due to the high covariances in JENDL-4.0 compared to those in ENDF/B-VII.1, in particular for the elastic cross section of the U-235 and of the fission for the U-238.

These results demonstrated that the covariances of most neutron reactions with the nuclei studied in this work require more investigation and re-estimation.

Acknowledgements

The work leading to this publication has been supported by the Radiations and Nuclear Systems Laboratory at Abdelmalek Essaadi University of Tetuan. Thank you to all the contributors of our laboratory team.

Author details

Mustapha Makhloul[1*], H. Boukhal[1], T. El Bardouni[1], E. Chakir[2], M. Kaddour[1] and S. Elouahdani[1]

1 Radiations and Nuclear Systems Laboratory, Faculty of Sciences of Tetuan, University Abdelmalek Essaadi, Morocco

2 SIMO Lab, Faculty of Sciences of Kenitra, Morocco

*Address all correspondence to: mustapha342011@hotmail.fr

References

[1] Cabellos O. Presentation and discussion of the UAM/exercise I-1b: "Pin-cell burn-up benchmark" with the hybrid method. Science and Technology of Nuclear Installations. 2013;**2013**:1-12. DOI: 10.1155/2013/790206

[2] Vibha V, Mukherjee S, Naik H, Parashari S, Makwana R, Suryanarayana SV. $^{238}U(n,\gamma)$ Reaction Cross-Section at the Neutron Energy 8.96 MeV; 2017

[3] Iwamoto O et al. Uranium-235 Capture Cross-Section in the keV to MeV Energy Region; 2011

[4] Palmiotti G et al. Combined use of integral experiments and covariance data. Nuclear Data Sheets. 2014;**118**: 596-636. DOI: 10.1016/j.nds.2014.04.145

[5] Otuka N, Nakagawa T, Shibata K. Uranium-235 neutron capture cross section at keV energies. Journal of Nuclear Science and Technology. 2007; **44**(6):815-818. DOI: 10.1080/18811248.2007.9711318

[6] Pelowitz DB et al. MCNP6 User's Manual. Los Alamos National Laboratory; 2013

[7] Macfarlane RE, Muir DW, Boicourt RM, Kahler AC. The NJOY Nuclear Data Processing System, Version 2012. Los Alamos National Laboratory (LANL); 2012

[8] Briggs JB. International handbook of evaluated criticality safety benchmark experiments. Nuclear Energy Agency, NEA/NSC/DOC (95). Exp. Needs Crit. Saf., vol. 3, Sep. 2004 [Online]. Available: https://www.oecd-nea.org/science/wpncs/icsbep/handbook.html

[9] Kaddour M. Analyse de sensibilité des problèmes de criticité et du coefficient de température aux données nucléaires: Contribution à l'amélioration des évaluations des sections efficaces et application à une maquette critique: le réacteur expérimentale EOLE; 2015

[10] Bowman SM. Experience with the SCALE Criticality Safety Cross-Section Libraries. Washington, DC: The Office: For sale by the U.S. G.P.O., Supt. of Docs.; 2000

[11] Broadhead BL, Rearden BT, Hopper CM, Wagschal JJ, Parks CV. Sensitivity and uncertainty based criticality safety validation techniques. Nuclear Science and Engineering. 2004; **146**(3):340-366

[12] Rochman D, Vasiliev A, Ferroukhi H, Zhu T, van der Marck SC, Koning AJ. Nuclear data uncertainty for criticality-safety: Monte Carlo vs. linear perturbation. Annals of Nuclear Energy. 2016;**92**:150-160. DOI: 10.1016/j.anucene.2016.01.042

[13] Sobes V, Leal L, Arbanas G, Forget B. Resonance parameter adjustment based on integral experiments. Nuclear Science and Engineering. 2016;**183**(3). DOI: 10.13182/NSE15-50

[14] Kiedrowski BC. 'MCNP6. 1 k-Eigenvalue sensitivity capability: a user's guide', Los Alamos National Laboratory (LANL), MCNP Documentation & Website; 2013. DOI: 10.13182/NSE10-22

[15] Chow ETY. An investigation of methods for neutron cross section error identification utilizing integral data [PhD thesis]. Georgia Institute of Technology; 1974

[16] Reupke WA. The consistency of differential and integral thermonuclear neutronics data [PhD thesis]. Georgia Institute of Technology; 1977

[17] Williams ML, Wiarda D, Ilas G, Marshall WJ, Rearden BT. Covariance

applications in criticality safety, light water reactor analysis, and spent fuel characterization. Nuclear Data Sheets. 2015;**123**:92-96

[18] Broadhead B, Hopper CM, Parks CV, Childs RL. Sensitivity and Uncertainty Analyses Applied to Criticality Safety Validation, methods development. United States: Oak Ridge National Lab; ORNL/TM–13692/V1. 1999

[19] Kuroi H, Mitani H. Adjustment to cross section data to fit integral experiments by least squares method. Journal of Nuclear Science and Technology. 1975;**12**(11):663-680

[20] Makhloul M, Boukhal H, El Bardouni T, Kaddour M, Chakir E, El Ouahdani S. ^{235}U elastic cross-section adjustment in criticality benchmarks – Comparison between JENDL-4.0 and ENDF/-VII.1. Annals of Nuclear Energy. 2018;**114**:541-550. DOI: 10.1016/j.anucene.2017.12.018

[21] Salvatores M et al. Methods and issues for the combined use of integral experiments and covariance data: Results of a NEA international collaborative study. Nuclear Data Sheets. 2014;**118**:38-71